Springer Tracts in Mechanical Engineering

T0172094

Springer Tracts in Mechanical Engineering (STME) publishes the latest developments in Mechanical Engineering - quickly, informally and with high quality. The intent is to cover all the main branches of mechanical engineering, both theoretical and applied, including:

- Engineering Design
- Machinery and Machine Elements
- Mechanical Structures and Stress Analysis
- Automotive Engineering
- Engine Technology
- Aerospace Technology and Astronautics
- Nanotechnology and Microengineering
- Control, Robotics, Mechatronics
- MEMS
- Theoretical and Applied Mechanics
- Dynamical Systems, Control
- Fluids Mechanics
- Engineering Thermodynamics, Heat and Mass Transfer
- Manufacturing
- Precision Engineering, Instrumentation, Measurement
- Materials Engineering
- Tribology and Surface Technology

Within the scope of the series are monographs, professional books or graduate textbooks, edited volumes as well as outstanding PhD theses and books purposely devoted to support education in mechanical engineering at graduate and post-graduate levels.

Indexed by SCOPUS, zbMATH, SCImago.

Please check our Lecture Notes in Mechanical Engineering at http://www.springer.com/series/11236 if you are interested in conference proceedings.

To submit a proposal or for further inquiries, please contact the Springer Editor **in your country**:

Dr. Mengchu Huang (China)
Email: mengchu.Huang@springer.com
Priya Vyas (India)
Email: priya.vyas@springer.com
Dr. Leontina Di Cecco (All other countries)
Email: leontina.dicecco@springer.com

More information about this series at http://www.springer.com/series/11693

Stefano Tornincasa

Technical Drawing
for Product Design

Mastering ISO GPS and ASME GD&T

 Springer

Stefano Tornincasa
DIGEP
Politecnico di Torino
Torino, Italy

ISSN 2195-9862 ISSN 2195-9870 (electronic)
Springer Tracts in Mechanical Engineering
ISBN 978-3-030-60856-9 ISBN 978-3-030-60854-5 (eBook)
https://doi.org/10.1007/978-3-030-60854-5

This Springer imprint is published by the registered company Springer Nature Switzerland AG
The registered company address is: Gewerbestrasse 11, 6330 Cham, Switzerland

Preface

Designers create perfect and ideal geometries through drawings or by means of Computer Aided Design systems, but unfortunately the real geometrical features of manufactured components are imperfect, in terms of form, size, orientation and location.

Therefore, technicians, designers and engineers need a symbolic language that allows them to define, in a complete, clear and unambiguous way, the admissible variations, with respect to the ideal geometries, in order to guarantee functionality and assemblability, and to turn inspection into a scientifically controllable process.

The *Geometric Product Specification* (**GPS**) and *Geometrical Dimensioning and Tolerancing* (**GD&T**) languages are the most powerful tools available to link the perfect geometrical world of models and drawings to the imperfect world of manufactured parts and assemblies.

This is a new, more complicated approach than the previous methodologies, but it offers the designer more opportunities and more powerful tools to define the expected functional requirements with the maximum allowed tolerance.

Torino, Italy   Stefano Tornincasa

About This Book

This book is intended for students, academics, designers, process engineers and CMM operators, and it has the main purpose of presenting the ISO GPS and the ASME GD&T rules and concepts. The GPS and GD&T languages are in fact the most powerful tools available to link the perfect geometrical world of models and drawings to the imperfect world of manufactured parts and assemblies. The topics that have been covered include a complete description of all the ISO GPS terminologies, datum systems, MMR and LMR requirements, inspection, and gauging principles.

Moreover, the differences between ISO GPS and the American ASME Y14.5 standards are shown as a guide and reference to help in the interpretation of drawings of the most common dimensioning and tolerancing specifications. The book may be used for engineering courses and for professional-grade programmes, and it has been designed to cover the fundamental geometric tolerancing applications as well as the more advanced ones. Academics and professionals alike will find it to be an excellent teaching and research aid, as well as an easy-to-use guide.

- Rules and concepts are explained with more than 400 original illustrations.
- The book is the result of a complex work of synthesis and elaboration of about 150 ISO standards and of the more recent ASME standards with the aim of clarifying technical rules and principles in order to document an industrial product in a univocal and rigorous manner.
- The latest changes and improvements of the ISO GPS and ASME Y14.5-2018 standards are presented.
- All the symbols that are used to interpret modern industrial technical drawings are described.
- This book represents an easy and indispensable guide for designers and professionals to clarify the concepts, rules and symbols pertaining to the complex world of technical product documentation.

Contents

About the Author

Prof. Stefano Tornincasa is full professor of Technical Drawing and Design Tools for Industrial Engineering at Politecnico di Torino. He has carried out research activities for over 30 years in the field of functional design and geometric tolerances and has published more than 200 national and international scientific papers. He was President of the ADM Improve Association (Innovative Methods in PROduct design and deVElopment) from 2011 to 2015.

He is co-author of the best-selling book on Industrial Technical Drawing, which is currently adopted in the design courses of most Italian universities.

Professor Tornincasa has conducted training courses on GD&T in many of the main manufacturing companies in Italy, and it is from this activity that he has derived his skill and experience in functional design.

His other research topics have been focused on product development, cycle innovation through digital models and virtual prototyping methodologies (PLM).

Chapter 1
Introducing GD&T and GPS

Abstract The technical product documentation that is currently drawn up in many companies is unfortunately still ambiguous and contains many errors, such as erroneous datums, imprecise or missing distances, as well as incongruent and difficult to check tolerances. For about 150 years, a tolerancing approach called "coordinate tolerancing" was the predominant tolerancing system used on engineering drawings. This methodology in fact no longer results to be the most suitable for the requirements of the modern global productive realty, in which companies, for both strategic and market reasons, frequently resort to suppliers and producers located in different countries, and it is therefore necessary to make use of communication means in which the transfer of information is both univocal and rigorous. GD&T or GPS is a symbolic language that is used to specify the limits of imperfection that can be tolerated in order to guarantee a correct assembly, as well as the univocal and repeatable functionality and control of the parts that have to be produced.

1.1 The Shortcoming of Traditional Coordinate Tolerancing

Geometric Dimensioning and Tolerancing (GD&T) or *Geometric Product Specification* (GPS) is a fundamental design tool that is used to clearly and unambiguously define the allowable limits of imperfection of as-produced parts, with the aim of guaranteeing assembly and functionality.

GD&T, or GPS, is a language that is made up of symbols which can be used to accurately control tolerances, but it also allows the maximum manufacturing flexibility and control of the costs to be achieved.

Real surfaces can in fact differ, to various extents, from the exact geometric form foreseen during the design phase, both as far as the geometric form is concerned and for the pre-established position with respect to other surfaces assumed as a reference, as a result of various factors (bending of the workpiece and of the tool during the tooling, vibrations of the machine, deformations during hardening, etc.).

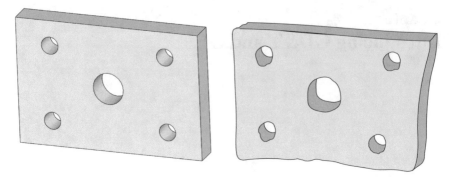

Fig. 1.1 The real surfaces of a constructed part may deviate to various extents from both the geometrically exact form indicated during the design phase and from the pre-established position with respect to surfaces or points assumed as a datum.

The workpiece in Fig. 1.1 is shown with the form errors conveniently enlarged for explanatory purposes, but it can also appear as such for the instruments that perform precise measurements (such as the modern *Coordinate Measuring Machines*, **CMM**).

Another example of form error is illustrated in Fig. 1.2: the sinuous trend of the workpiece or not detected by the measurement gauge, which locally ascertains, section by section, the values of the diameters that fall within the dimensional tolerance limits. However, it would be difficult to insert the workpiece easily into an axially extended hole, because the spatial dimension of the pin would be greater than that foreseen for the tolerance.

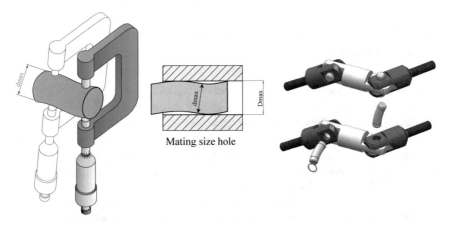

Mating size hole

Fig. 1.2 Evaluating the mating conditions by only referring to linear distances leads to the effects of the form being neglected: a smooth f/H shaft-hole mating is in practice forced if the roundness and straightness conditions are not adequate. In fact, the measured maximum dimension, dmax, that conforms to the requested tolerance is less than the Dmax dimension that corresponds to the actual mating dimensions of the pin. Accordingly, it is not possible to fit the hinge pin with the hole because of the error in the straightness of the axis

It is hence necessary to set adequate constraints on the straightness of the axis and on the roundness of the surface. It therefore seems evident that, when establishing the errors that are acceptable in the construction of a part, the form and dimensions should be evaluated in light of the functional requirements. Above all, as the complexity of the designed objects increases, as a result of the numerous technological processes that are now utilised and because of the necessity of guaranteeing quality through careful inspections and verifications, the information that is derived from the project, and therefore from the design of the component that has to be produced, should refer to all the sectors involved as much as possible. The drawings of the workpiece, as requested for the process, should not be ambiguous; the function of the parts, as well as the way of obtaining and controlling them should be completely included and reported.

In short, each feature of a workpiece **should be fully defined in terms of dimension, position, orientation and form**.

The tolerances, whether geometrical or dimensional, should be essential parts of the design process right from the beginning, and not just an accessory that has to be added only once the design has been completed.

One procedure that can be adopted to satisfy these requirements is the **functional dimensioning method**, which can be defined as a way of specifying the geometric functionalities and the functional relationships that exist among the form characteristics, in a project or design, in order to obtain the most valid production from both a qualitative and an economic point of view.

The components that have to be controlled and measured can be specificated through this method; in this way, the intentions of the designer are respected, and the manufacturer is able to choose the most appropriate manufacturing procedures.

Computer Aided Design (**CAD**) tools, which have the purpose of generating, manipulating and reporting the geometry of the parts that have to be produced, are currently used for design purposes, but they are not sufficient on their own to document an industrial product in an effective way.

In order to communicate a precise and rigorous description of a part, not only should the dimensions be indicated in the engineering drawing, but also the admissible error, in terms of **size**, **location**, **orientation** and **form**.

The term *"Geometric Dimensioning and Tolerancing"* (GD&T) is often used to characterise a functional design. GD&T is a language made up of symbols that are used to specify the limits of imperfection that can be tolerated in order to guarantee a correct assembly, as well as the univocal and repeatable functionality and control of the parts that have to be produced. Moreover, GD&T is a *tool that allows an imperfect geometry to be managed "perfectly"*.

The GD&T method was developed and extended during the Second World War by Great Britain and the United States. The first publication concerning a standard, pertaining to the basic concepts of form and position, was published in Great Britain in 1948.[1] In 1940, in the United States, Chevrolet printed a publication concerning position tolerances, and this was followed, in 1945, by a manual, which was published by

---

[1] This history is detailed in Sect. 4.1.1.

the American Military administration, in which symbols were introduced to specify form and position tolerances. Over the following years, the ASA and SAE associations published their own standards, which, in 1966, took on the present forms, although they have periodically been updated (the most recent being in 2018).

About 50 years ago, some companies in Europe began to introduce a few indications relative to admissible geometric errors, and the first mention of this topic appeared in texts on comparative design. In 1969, an ISO recommendation project introduced, at a normative level, the symbols that are used today. However, geometric tolerance was long considered an option that was only to be used in particular and rare occasions, that is, when the dimensional tolerances (which were generally considered suitable if accompanied by a correct execution) were not considered sufficient to define the exact form of a component.

A different approach to tolerances, which couples geometric dimensioning with the functionality of a workpiece, is currently being developed, with the aim of obtaining a shared, clear and comprehensive language for all the production sectors interested in a particular project, and of reporting the finished conditions of the objects that have to be produced in a univocal and faithful way, with the intention of advantaging functionality and reducing costs. As already mentioned, this is not a recent method, but it has still not been fully exploited in many European countries, where the benefits that could be achieved, in practical terms, on the highly competitive markets of today, are often disregarded.

Unfortunately, the technical product documentation that is currently drawn up in many companies is still ambiguous and often contains many errors, such as erroneous datums, imprecise or missing distances, as well as incongruent and difficult to check tolerances. These documents were in fact first conceived considering a series of standards that originated in the technical industrial context of the last century, that is, by referring to the then most widespread measurement instruments, such as gauges and micrometres, in which the measurements were made and expressed as the distance between two points. The most frequently used measurement method was that of coordinate dimensioning, where only the dimensional tolerances (that is, the "*plus or minus*" system), which could give rise to manufacturing and control ambiguity, were used.

This methodology is in fact no longer results to be the most suitable for the requirements of the modern global productive realty, in which companies, for both strategic and market reasons, frequently resort to suppliers and producers who are located in different countries, and it is therefore necessary to make use of communication means in which the transfer of information is both univocal and rigorous.

1.2   Traditional Dimensioning Example

Let us consider the drawing in Fig. 1.3 of a plate with 4 holes, for which the assembly with fixed fasteners is demonstrated as an example; the traditional measurement method has been applied to the plate, that is, with the 4 holes being located according to the dimensional tolerances.

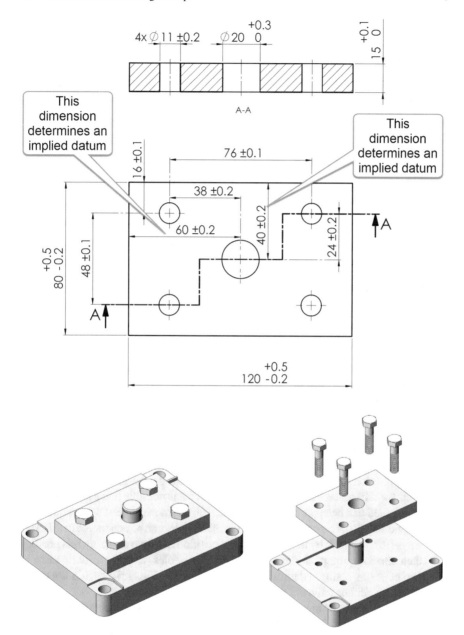

Fig. 1.3 Traditional dimensioning of a plate with 4 holes, for which the functional assembly conditions are shown (coherence of the profiles, perpendicularity of the profile, location of the holes, implied datums)

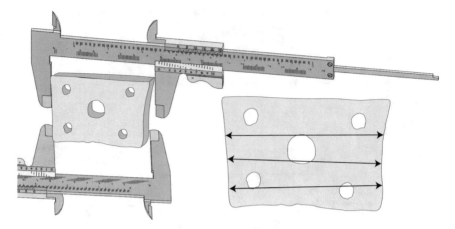

Fig. 1.4   If the measurement of the dimensions starts from the edges, which are not orientated at 90°, as shown in the drawing, how should the workpiece be arranged exactly to control the dimensional tolerance? Furthermore, as the measurement is not the individual distance between pairs of opposite points, how are the opposite points determined?

An observation of the thus defined workpiece leads to the following considerations and questions:

(1)   Are the edges of the workpiece located with respect to the holes or are the holes located with respect to the edges [1]?

(2)   If the distances are considered to originate at the edges, which are not orientated at 90°, as shown in the drawing (Fig. 1.4), how should the workpiece be arranged exactly to control the dimensional tolerance? Furthermore, the measurement is not just the individual distance between pairs of opposite points.

(3)   In coordinate tolerancing, since datums are implied and they depend on the interpretation of the metrologists, the placement on the gauging surface may produce different measurement results. Hence it is possible that a good part may be rejected or a bad part may be accepted. Furthermore, if the datum sequence is not indicated, the control of the location of the holes can be conducted in different ways (for example, by first checking the position of a hole and from this position then checking the positions of the subsequent holes, or by considering one or two edges as a single datum for all the holes, see Fig. 1.5): such a control is therefore neither univocal nor repeatable.

(4)   The cross-sectional area of the tolerance zone for the position control of the hole axis is *square*, while the form of the hole is *round*; therefore, the tolerance area does not mirror the form of the hole that it should protect. Moreover, starting from the theoretical position (point 0 in Fig. 1.6), various possible limit positions of the axis are possible, depending on the radial directions (a possible error of 0.28 mm can be reached along the diagonal of the square, that is, a greater error than the tolerance of 0.2 indicated on the drawing).

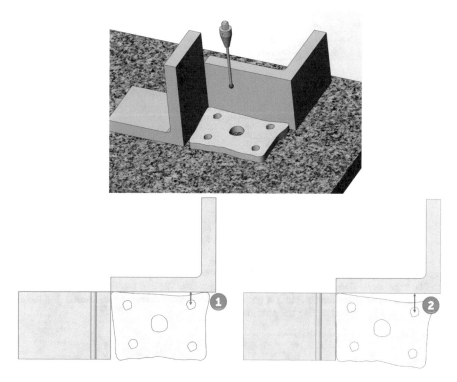

Fig. 1.5 Without any indication of the datums, the control of the position of the holes is not univocal or repeatable (measurements 1 and 2 give different results, according to the support system of the part used during the verification)

(5) A series of measurements can lead to an accumulation of the error pertaining to the position of the holes; Fig. 1.7 shows that it is important to report, in a clear way, the control method that is used for the positioning of a hole, which, if left to the discretion of an inspector, could lead to a variety of very different results.

(6) Finally, the workpiece may be rejected during the control because it does not conform with the tolerances prescribed during the design phase, even though it can be coupled and it is functional.

Inconveniencies of this type are more common than one might think, and they may lead to negative effects on the cost and times necessary for the construction of a component.

In short, the technical product documentation of a product without functional geometric tolerances and without any indication of datums is generally incomplete, and therefore not unambiguously interpretable. Such an incomplete, ambiguous tolerancing of components on engineering drawings leads not only *to increased production and inspection costs* but also to *an incalculable liability risk* in the case of legal disputes.

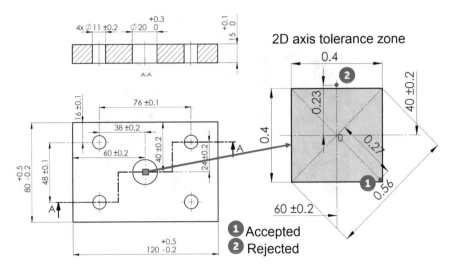

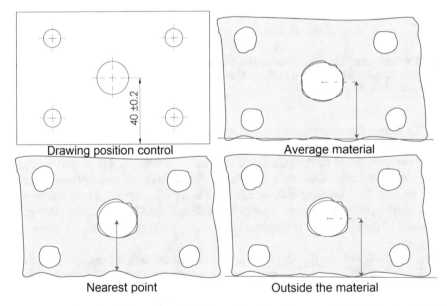

Fig. 1.6 Starting from the theoretical position (point 0), the limit positions allowed for the axis may be larger or smaller, depending on the radial directions. The axis position of the hole at point 1 is accepted during the control, but if it occurs at point 2, the workpiece is discarded. But which of the two holes is closest to the ideal position?

Fig. 1.7 The control of the position of a hole may be achieved by resorting to of the use of ideal circles and surfaces (known as *associated geometry* in the ISO system). The three verification methodologies shown in the figures could furnish contrasting results

Reference

1. Krulikowski A (1997) Fundamentals of GD&T self-study workbook, 2nd edn. Effective Training Inc.

Chapter 2
The Geometrical Product Specification (GPS) Language

Abstract The main concepts proposed in this chapter have the aim of expressing the fundamental rules on which the geometrical specification of workpieces can be based through a global approach that includes all the geometrical tools needed for GPS. Indeed, in an increased globalisation market environment, the exchange of technical product information and the need to unambiguously express the geometry of mechanical workpieces, are of great importance. The symbols, terms, and rules of the GPS language that are given in the ISO 17450, ISO 1101 and ISO 14660 standards are presented in this chapter through tools and concepts that allow an engineer to perfectly specify the imperfect geometry of a component, and to understand the impact of the drawing specifications on inspection.

2.1 General Concepts

The current industrial situation is characterised more and more by a continuous evolution towards increasingly dynamic interaction models between clients and suppliers, thereby putting traditional technical communication methodologies under greater pressure.

In a market environment of increased globalisation, the exchange of technical product information and the need to unambiguously express the geometry of mechanical workpieces is of great importance. Consequently, the codification associated with the functional geometrical variations of the macro- and micro-geometries of workpiece specifications needs to be unambiguous and complete. An ever increasing requirement of accuracy has been observed in the description and in the interpretation of the functional requirements, and, consequently in the drawing up of the technical designs and documents in the mechanical subcontracting sector, which must be coherent and complete and able to adequately support the co-design and outsourcing requirements of a production.

In addition, the rapid development of the coordinate measuring technique indicates that some specifications based on measurements with gauges and calibres, that were used in the last quarter of the XXth century, may be interpreted differently and produce different measurement results.

S. Tornincasa, *Technical Drawing for Product Design*, Springer Tracts
in Mechanical Engineering, https://doi.org/10.1007/978-3-030-60854-5_2

Today, the importance of international standards in the technical documentation field is growing at the same rate as the globalisation of production; a simple, clear, univocal and concise three-dimensional description of the designed components is therefore necessary.

For this reason, a remarkable effort is under way to develop a coherent and innovative management scheme of geometric tolerances, in order to obtain a better definition of the correlation between the functional requirements, geometrical specifications and relative control procedures, which can be summarised as the **Geometrical Product Specification**—*GPS* and **Geometric Dimensioning and Tolerancing**—*GD&T* principles, and which, if implemented correctly and coherently, allow the drawbacks of the present methodologies to be overcome and intra and inter-company communication to be revolutionised.

Although the discussion of the new rules pertaining to the GD&T methodology is put off until the next section, it is here considered opportune to recall some basic definitions in the context of what is now defined, in the rules and in practice, as GPS, that is, Geometrical Product Specification.

Geometrical product specifications (GPS) defines a number of new concepts that have the potential of revolutionising the specification and verification domain, thereby enabling a designer to completely and unambiguously express the functional requirements in the technical documentation of products.

The basic tools of the GPS language are given in the ISO 17450 and ISO 14660 standards. According to the ISO 14660/1 standard, the geometrical features can be found in three domains:

- the **specification domain**, where several representations of the future workpiece are imaged by the designer;
- the **workpiece domain**, that is, the physical world domain;
- the **inspection domain**, where a representation of a given workpiece is used through the sampling of the workpiece with measuring instruments.

Table 2.1 shows some basic concepts of ISO 14660/1, which are often referred to in the ISO 17450/1 standard as the definitions of **integral, associated** and **derived** features.

An *integral* feature is a geometrical feature that pertains to the real surface of a workpiece or to a surface model. An *associated* feature is an ideal feature that is established from a non-ideal surface model or from a real feature through an association operation. A *derived* feature is a geometrical feature that does not exist physically on the real surface of a workpiece, but which can be established from a nominal feature, an associated feature, or an extracted feature. The centre of a sphere is a derived feature obtained from a sphere, which is itself an integral feature. The median line of a cylinder is a derived feature that is obtained from the cylindrical surface, which is an integral feature (Fig. 2.1). The axis of a nominal cylinder is a nominal derived feature.

In metrology, through the use of a CMM measuring machine, the *extracted feature* (integral or derived, that is, an approximated representation of the real feature, which

Table 2.1 Relationship of the three domains and the feature types

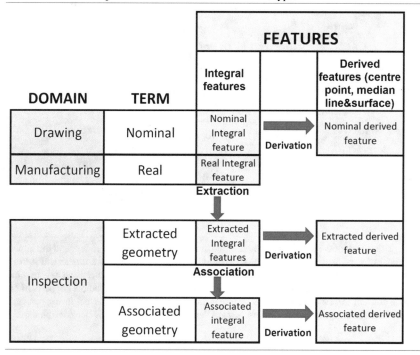

		FEATURES		
DOMAIN	**TERM**	Integral features		Derived features (centre point, median line&surface)
Drawing	Nominal	Nominal Integral feature	Derivation	Nominal derived feature
Manufacturing	Real	Real Integral feature		
		Extraction		
Inspection	Extracted geometry	Extracted Integral features	Derivation	Extracted derived feature
		Association		
	Associated geometry	Associated integral feature	Derivation	Associated derived feature

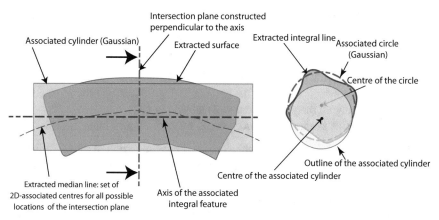

Associated cylinder (Gaussian)

Intersection plane constructed perpendicular to the axis

Extracted surface

Extracted integral line Associated circle (Gaussian)

Centre of the circle

Outline of the associated cylinder

Centre of the associated cylinder

Extracted median line: set of 2D-associated centres for all possible locations of the intersection plane

Axis of the associated integral feature

Fig. 2.1 Illustration of the process adopted to build an extracted median line: an extracted median line is a set of 2D-associated centres

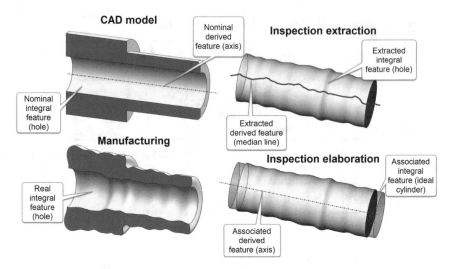

Fig. 2.2 The ISO 14660/1 standard provides terms that allow an engineer to understand the impact of the drawing specifications on inspection. A nominal integral feature is a theoretically exact feature that has been defined in a technical drawing. A nominal derived feature is an axis that has been derived from one or more integral features. Extracted and associated features are parts of the inspection domain. An associated integral feature is an integral feature of a perfect form associated with the extracted integral feature. An associated derived feature is an axis or centerplane of a perfect form.

is acquired by extracting a finite number of points) is obtained from the real integral feature (Fig. 2.2). A perfect associated feature (a cylinder or the derived axis, which can be used, for example, as a datum) can thus be obtained from the extracted features.

The "world" of manufacturing disappears in ISO 17450/1:2011, where it is replaced by the "world" of **specification**. ISO 17450 is in fact aimed at expressing the fundamental concepts on which the geometrical specification of workpieces can be based, including all the geometrical tools needed in GPS.

According to this standard, the geometrical features exist in three "worlds":

(1) the *nominal* definition world, where an ideal representation of the workpiece is defined by a designer with a perfect form, i.e. with the shape and dimensions necessary to meet the functional requirements (Fig. 2.3). This workpiece is called the "**nominal model**", and it is impossible to produce or inspect since each manufacturing or measuring process has its own variability or uncertainty.

(2) the *specification* world, where a designer, from the nominal geometry, imagines a model of this real surface, which represents the variations that could be expected on the real surface of the workpiece. This model, which represents the imperfect geometry of the workpiece, is called the "**non-ideal surface model**". The non-ideal surface model is used to simulate variations of the surface at a conceptual level (see Fig. 2.4), thereby optimising the maximum permissible limit values for which the function is downgraded but still ensured. These maximum permissible limit values *define the tolerances of each characteristic of the workpiece.*

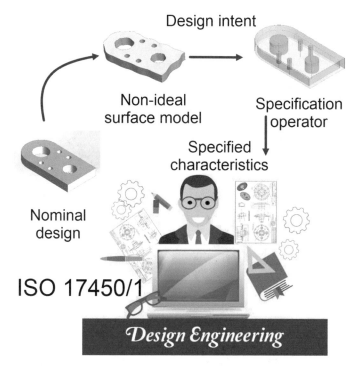

Fig. 2.3 The *specification* world, where a designer, from the nominal geometry, imagines a model of this real surface, which represents the variations that could be expected on the real surface of the workpiece

(3) the *verification* world, where a metrologist defines the sequence of operations that will be used during the measuring process. The metrologist reads the specification of the non-ideal surface model, in order to define the individual steps of the verification plan, (i. e. the mathematical and physical operations of the sampling of the workpiece). The measurement, through the so-called **duality** principle defined in ISO 17450–1, can mirror the specification operation for operation. **Conformance** is then determined by comparing the specified characteristics with the result of the measurement (see Fig. 2.5).

The specification process is the first process to take place in the definition of a product or a system. The purpose of this process is to translate the designer's intent into a requirement or requirements for specific GPS characteristics. The designer is responsible for the specification process, which comprises ensuring the following steps (Fig. 2.6):

(a) *feature functionality*, i. e. the desired design intent of the GPS specification;
(b) *GPS specification*, which consists of a number of GPS specification elements each of which controls one or more specification operations;

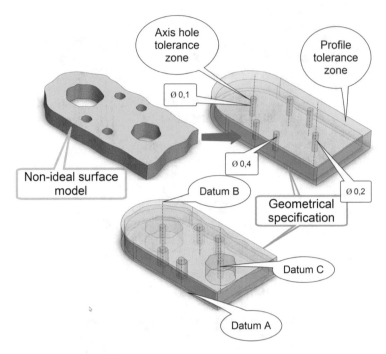

Fig. 2.4 The non-ideal surface model is used to simulate variations of the surface at a conceptual level, thus optimising the maximum permissible limit values for which the function is downgraded but still ensured

(c) *specification operations*, which are defined as an ordered set of operations and can be thought of as a *virtual* measurement instruction, where each operation and the parameters that define such an operation are steps in the measuring process.

A specification operator is necessary to define, for example, a possible specific "diameter" in a cylinder (two-point diameter, minimum circumscribed circle diameter, maximum inscribed circle diameter, least squares circle diameter, etc.), instead of the generic "diameter" concept. A specification involves expressing the field of permissible deviations of a characteristic of a workpiece as the permissible limits. There are two ways of specifying the permissible limits: by considering the dimension or by specifying the zone that limits the permissible deviation of a non-ideal feature inside a space.

This is a new, *more complicated approach* than the previous methodologies, but it offers the designer *more opportunities and more powerful tools* to define the expected functional requirements *with the maximum allowed tolerance*.

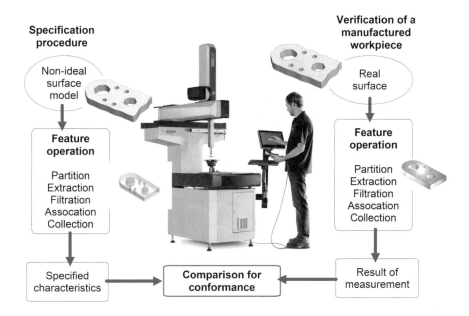

Fig. 2.5 The GPS **duality** principle pertaining to the specification and verification procedures: the metrologist reads the specification of the non-ideal surface model, in order to define the individual steps of the verification plan. Conformance is then determined by comparing the specified characteristics with the measurement result

2.2 Classification of Geometric Tolerances

It is necessary to reiterate that dimensional errors (pertaining to the length and diameters) and form errors can be encountered in the manufacturing of a component, but also in the orientation and/or location between two features of a part, such as in the previous examples; geometrical type tolerances can in fact be divided into four categories:

(1) **Form** tolerances, which establish the variation limits of a surface or a single feature of the ideal form indicated in the design. The form of an isolated feature is correct when the distance of each of the points of an ideal geometric surface, which are in contact with the datum, is equal to or less than the indicated tolerance; it should be pointed out that the tolerances on the profiles constitute a group on their own, in that they not only establish the limits of variation of the absolute form, but also the location and orientation of a surface or of any line, with respect to a possible datum.

(2) **Orientation** tolerances, which establish the limits of variation of a surface of a single feature with respect to one or more features assumed as a datum. The datum feature may be an already existing feature of the part, and its form should be sufficiently precise for it to be used as such.

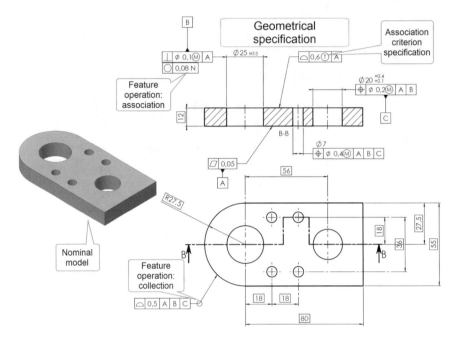

Fig. 2.6 The specification operations, defined as an ordered set of operations can be thought of as a virtual measurement instruction, where each operation and the parameters that define such an operation are steps in the measuring process. The permissible limits are specified by the application of the tolerance zones according to the ISO 1101:2017 standard

(3) **Location** tolerances, which establish the limits of variation of a surface or a single feature with respect to an ideal location, as established in the design process, and to one or more features assumed as a datum.

(4) **Run-out** tolerances, which establish the limits of variation of a surface or single feature with respect to a form and a location, established in the design phase, during the rotation of a piece around a reference feature.

The chart in Table 2.2 shows the characteristic geometrical symbols used for geometrical tolerancing and the four tolerance types into which these symbols are divided. The symbols are explained in more detail later on.

As can be seen from the table, some non-associable tolerances (for instance, planarity, roundness, etc.) do not refer to other features of a part, taken as a datum; some geometrical characteristics require a datum (such as parallelism or the location of a hole) and other tolerances that may or may not be associated with another feature, such as tolerances on profiles.

In other words, while a form error concerns an isolated feature (such as a flat surface), an orientation or location error is associated with another feature of the workpiece, which is known as the datum feature (more details of which will be presented in the next sections, together with details on how such a feature should be

Table 2.2 Classification, types and symbols of geometrical tolerances

Tolerances	Geometric attributes	Datum	Geometrical characteristic	Symbol
Form	Does not control size, orientation or location	NO	Straightness	
			Flatness	
			Roundness[2]	
			Cylindricity	
Profile	Form, size, orientation and location control	YES/NO	Line profile[3]	
			Surface profile[4]	
Orientation	Does not control size or location[5]	YES	Parallelism	
			Perpendicularity	
			Angularity	
Location	Does not control form or size	YES	Position	
			Concentricity[6] and coaxiality	
			Symmetry[6]	
Run-out[7]	Does not control size	YES	Circular run-out	
			Total run-out	

[2] Circularity in ASME

[3] Profile of a line in ASME

[4] Profile of a surface in ASME

[5] The orientation controls form when applied to a flat surface

[6] Both concentricity and symmetry have been removed from the ASME Y14.5:2018

[7] Runout in ASME

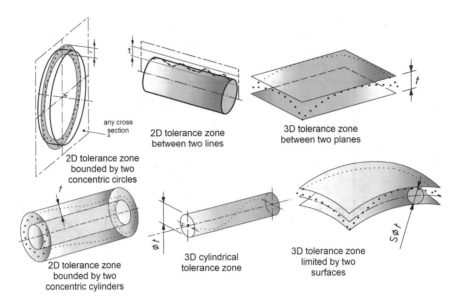

Fig. 2.7 Some typical forms of three-dimensional tolerance zones

indicated). In short, a geometrical tolerance defines a space (either bi-dimensional or three-dimensional) within which the feature that has to be controlled must remain. For example, a feature of a plane, for which one wishes to control the straightness, should remain within the area defined by two straight parallel lines at the same distance as the tolerance value; the tolerance zone in the 3D space should be a cylinder with a perfectly straight axis and generatrix, and with the diameter equal to the tolerance value.

Some typical forms of tri-dimensional tolerance zones are visible in Fig. 2.7, together with indications on the relative dimensions.

2.3 The Specification World: Types and Symbols of Geometrical Tolerances

2.3.1 ISO Tolerance Indicator

According to the aforementioned standard, that is, ISO 1101, geometrical tolerances should be indicated on drawings by means of a rectangular frame (ISO: **tolerance indicator**, ASME: **feature control frame**) divided into two or more compartments (Fig. 2.8). The compartments should contain, from left to right, the following indications in the same order:

(1) the *symbol* of the geometrical tolerance, according to Table 2.2

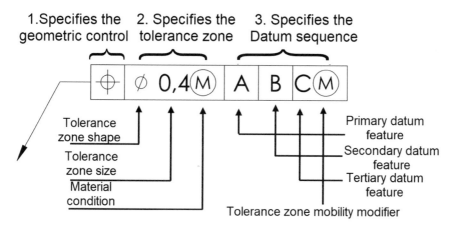

Fig. 2.8   The tolerance indicator decoding process

(2)   the *total tolerance value* of the measurement unit used for the linear dimensions. This value is preceded by the sign Ø if the tolerance zone is round or cylindrical; another indication can be given in this compartment, that is, a capital letter inserted into a circle, which is used to prescribe the so-called "modifiers", such as, for example, the application of the requirement of the maximum material, and so on, which modify the tolerance value;

(3)   the letter or letters that identify the *datum features*, whenever necessary, which may be followed by the indication of modifiers.

Any necessary annotations may be written either above or close to the tolerance indicator (for example, "6 x", as in Fig. 2.9) or joined to the frame by means of a leader line. If more than one geometrical tolerance is indicated on the same feature, the indications should be reported in overlapping compartments, as shown in Fig. 2.10.

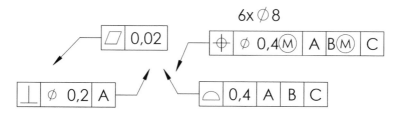

Fig. 2.9   Examples of geometrical specifications with a tolerance indicator. The geometrical specification indication should be connected to the leader line by a reference line. The reference line should be attached to the mid-point of the left-hand side or the right-hand side of the tolerance indicator

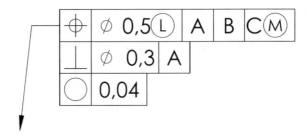

Fig. 2.10 If more than one geometrical tolerance is indicated on the same feature, the indications should be reported in overlapping compartments. In this case, it is recommended that the tolerance indicators are arranged so that the tolerance values are shown in descending order, from top to bottom

2.3.2 Drawing Geometrical Specifications

The tolerance frame is connected to a toleranced feature by a leader line, with an arrow at the end:

(a) on the contour line of the feature or on an extension line of the contour (but clearly separate from a measurement line), if the tolerance is applied to a line or surface (Fig. 2.11); it is also possible to use a reference line that points directly to the surface.

(b) According to ISO/CD 16792:2011, pertaining to the definition of 3D digital products, leader lines that have the aim of representing line elements should terminate with an arrowhead, see Fig. 2.12. When an indicated element is a surface, the leader line should terminate with a dot within the bounds of the surface.

(c) as an extension of the dimension line, if the tolerance applies to a derived feature (axis, median line, median surface or a centre point, Fig. 2.13).

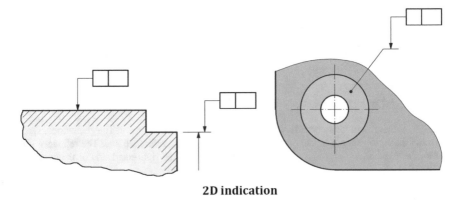

2D indication

Fig. 2.11 Indication of an integral feature specification

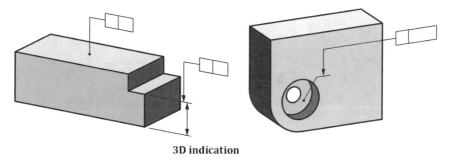

3D indication

Fig. 2.12 In 3D annotation, when an indicated element is a surface, the leader line should terminate with a dot within the bounds of the surface

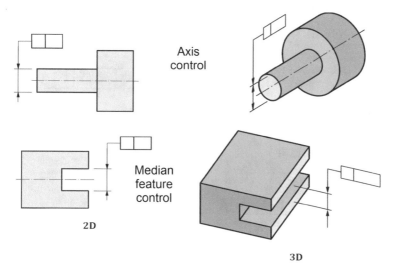

Axis control

Median feature control

2D

3D

Fig. 2.13 Indication of a derived feature specification

The direct joining of a reference element to the frame through a leader line, with the omission of the datum letter, should be avoided. The practice of directly connecting a tolerance indicator with a leader line terminating with an arrow directly to the axis has been deprecated (Fig. 2.14); see Fig. 2.13 for the preferred indications.

The letter that indicates the datum feature should be reported in the frame, as shown in Fig. 2.15. Although a single datum feature is identified by just one single capital letter (Fig. 2.15a), a common datum, established from two features, should be identified by two different letters separated by a hyphen (Fig. 2.15b).

As will be seen later on, the datum system is not only useful to establish the functional relationships between several features, but also to indicate the sequence that has to be followed to control the geometrical tolerance of a part; this means that,

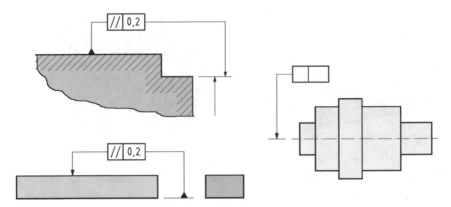

Fig. 2.14 It was former practice to connect the tolerance indicator directly to the datum feature by means of a leader line or to connect the tolerance indicator by a leader line terminating with an arrow directly to the axis

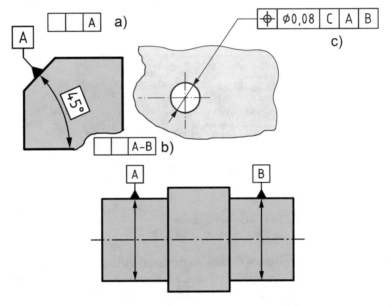

Fig. 2.15 Indication of a single datum feature (**a**), a common datum established from two features (**b**) and three datums, in priority order, from left to right (**c**)

in the case of several datums, the relative letters should be indicated in consecutive compartments in the frame (Fig. 2.15c), in priority order from left to right.

If the tolerance is applicable to a limited length, which is not defined as a location, the value of this length should be added to the tolerance value and separated from it by an oblique stroke. If a more restrictive tolerance on a limited length is added

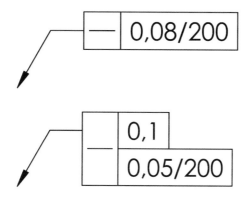

Fig. 2.16 Indication of restrictive specifications. If two or more specifications of the same feature have to be indicated, they may be combined. they may be combined

to another tolerance on a feature, the tolerance indicator should be placed below the other, as indicated in Fig. 2.16.

If the tolerance (or datum) only has to be applied to a restricted area of a feature, it should be identified (with locations and dimensions defined according to **Theoretically Exact Dimensions** (TED),[1] as indicated in Fig. 2.17, using a chain line (line 4.2 in ISO 128–24).

Inserting a circle, in correspondence to the junction of the leader line of the tolerance (Fig. 2.18), indicates that the geometric control applies to *the entire external boundary* (**all around**), while two circles indicate that a profile tolerance, or other specification, applies over *the entire three-dimensional profile* of a part (**all over**). An "all around" or "all over" symbol should always be combined with an SZ (*separate zones*), CZ (*combined zone*) or UF (*united feature*) specification element, except for when the referenced datum system locks all non-redundant degrees of freedom (see the next section).

2.3.3 Additional Symbols of the ISO 1101 and ISO 5458 Standards

The ISO 1101:2017 standard has defined a new symbol to specify the derived features of a revolute surface (axis or median line), that is, the modifier Ⓐ (median feature) is placed in the tolerance section of the tolerance indicator. In this case, the leader line does not have to terminate at the dimension line, but can terminate with a dot on the integral feature, an arrow on the outline or an extension line, as shown in Fig. 2.19.

Another useful indication that is used to specify the constraints on the tolerance zone is the **Combined Zone**, which is indicated as **CZ**, and which is used to control

---

[1]Basic dimension in ASME.

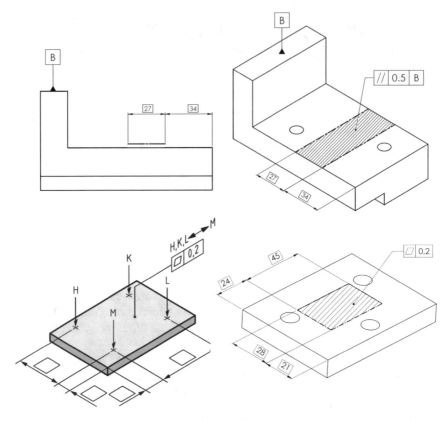

Fig. 2.17   Indication of a restrictive area toleranced feature

the tolerance zone applied to several surfaces with the aim of establishing a common tolerance zone that can be applied simultaneously to the indicated surfaces (Fig. 2.20). In this case, all the relative tolerance zones should be individually constrained, as far as the location and orientation between each other are concerned, even through the use of either explicit or implicit theoretically exact dimensions (TED) (Fig. 2.21).

2.3.3.1   Indications Adjacent to the Tolerance Indicator

A geometrical specification indication consists of a tolerance indicator with some optional adjacent indications. When there is only one tolerance indicator, indications in the upper/lower adjacent indication areas and in the in-line adjacent indication area mean the same. In this case, only one adjacent indication area should be used and it is preferable to use the upper adjacent indication area, if possible (Fig. 2.22).

   Table 2.3 shows other additional symbols (plane and feature indicators), that can be used to control the geometrical errors and allow the verification methodology to be

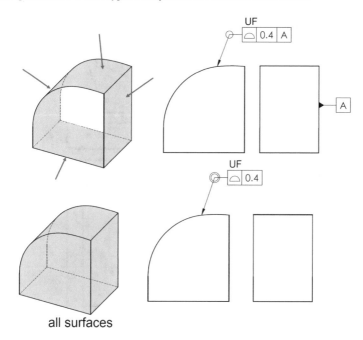

all surfaces

Fig. 2.18 All around specification: a geometrical specification is applied to the outlines of the cross-sections or when it is applied to all features represented by a closed outline. The all over symbol indicates that a geometrical specification is applied to all integral features of a workpiece

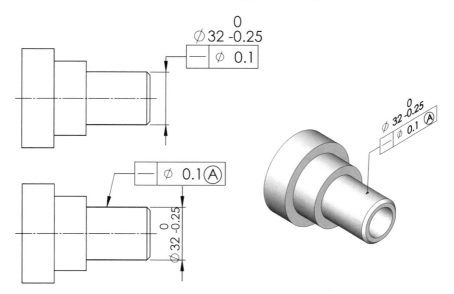

Fig. 2.19 In the case of a revolute, the derived median line can be indicated by a modifier (median feature) placed in a specific section of the tolerance indicator. In this case, the leader line does not have to terminate on the dimension line

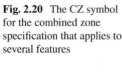

Fig. 2.20 The CZ symbol for the combined zone specification that applies to several features

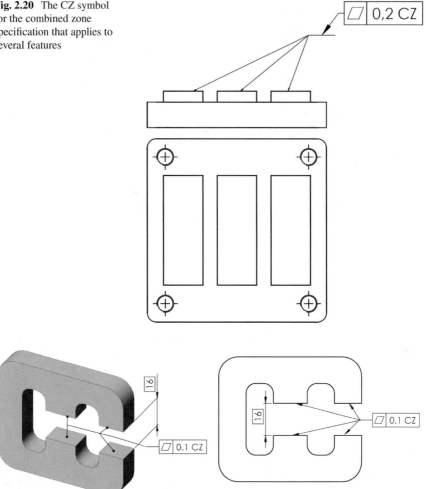

Fig. 2.21 All the related individual tolerance zones should be constrained in location and in orientation between each other using either explicit theoretically exact dimensions (TED), or implicit TEDs

specified in an unambiguous manner. Intersection plane, orientation plane, direction feature and collection plane indicators can be indicated in the in-line adjacent to the tolerance indicator.

For example, the use of the **intersection plane** indications (specified through the use of an intersection plane indicator placed, as an extension, to the right of the tolerance indicator) makes the identification of the bi-dimensional tolerance zone possible, regardless of which projection is used (Fig. 2.23), and it is therefore very useful for 3D dimensioning (Fig. 2.24).

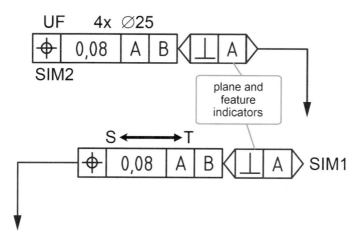

Fig. 2.22 Indications adjacent to the tolerance indicator

Table 2.3 Roles and symbols that can be used with intersection plane, orientation plane, direction feature and collection plane indicators

Name	Indicator	Symbols of the first compartment	Role
Intersection plane	◁ // B	⊥ // ∠ ≡	To identify the orientation of line requirements
Orientation plane	◁ // B ▷	⊥ // ∠	To control the orientation of a tolerance zone
Direction feature	◄ // C	⊥ // ∠ ↑	To indicate the direction of the width of the tolerance zone
Collection plane	◯ // A	⊥ // ∠ ≡	To indicate a family of parallel planes identifying the features covered by the "all around" indication

   The letter that identifies the datum used to establish the intersection plane is placed in the second section of the intersection plane indicator. The symmetry symbol is used to indicate that the intersection plane includes the datum.
   Figure 2.25 illustrates the use of a straightness tolerance on a flat surface as indicated in the ASME Y14.5:2018 standard. The intersection plane is not used, but the direction of the prescribed straightness tolerance is indicated by means of an orthographic view or by a line as a supplemental geometry in the 3D model.

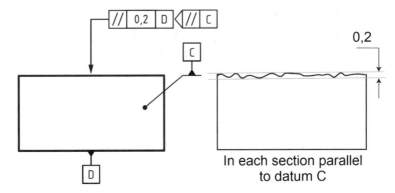

Fig. 2.23 Specification using an intersection plane indicator. The toleranced feature is made up of all the lines of a feature in a given direction, and the intersection plane is used to identify the orientation of the line requirements

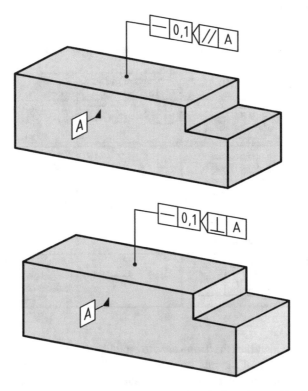

Fig. 2.24 In 3D annotation, the intersection plane should be indicated to avoid misinterpretation of the toleranced feature

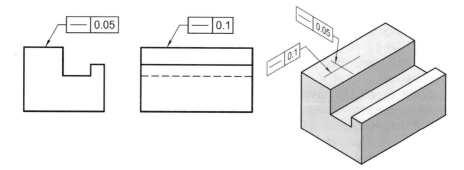

Fig. 2.25 Straightness applied to control line elements in multiple directions on a flat surface in the ASME Y14.5:2018 standard; the direction of the prescribed straightness tolerance is indicated by two lines as a supplemental geometry in the 3D model

The **orientation plane** indicator controls both the orientation of the planes that limit the tolerance zone (directly, by means of the datum and the symbol in the indicator) and the orientation of the width of the tolerance zone (indirectly, perpendicular to the planes), or the orientation of the axis for a cylindrical tolerance zone. A typical application used to control the angle of an axis is shown in Fig. 2.26, as an alternative

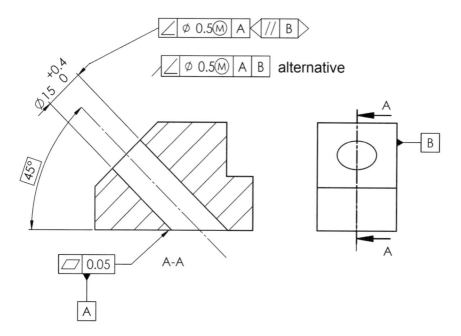

Fig. 2.26 The cylindrical tolerance zone is parallel to datum plane B and inclined at the specified angle to datum plane A; the orientation plane is used to avoid the presence of two datum features in the tolerance indicator

to the previous methodology, which foresaw the indication of two datums.

The new standard, that is, standard 1101 of 2017, has also clarified a misunderstanding that had existed for some time pertaining to the previous versions and which has finally been resolved, that is, the *indication of the direction of the tolerance zone of roundness*.

In fact, the old 1101 standard foresaw, in Sect. 8.1, that "*the amplitude of all the roundness tolerance zones, if not otherwise expressed, shall always be developed in an orthogonal direction to the surface of the feature that had to be controlled*". However, just a few lines later, at the end of the section, two lines, which stated "*in the case of roundness, the amplitude of the tolerance zone shall be applied in an orthogonal direction to the nominal axis of the piece*" appeared in a cryptic way.

Instead, in the new version, in Sect. 7.1, the new standard specifically states, notwithstanding what is prescribed by default, that the new "**direction feature**" symbols should be used for surfaces that are not round or cylindrical to specify how the tolerance is applied. Therefore, a direction indicator should always be indicated for the roundness of conical features. In Fig. 2.27, for both the cylindrical and conical surfaces, the extracted circumferential line, in any cross-section of the surfaces, should be contained between two coplanar concentric circles, with a difference in radii of 0,03. This is the default value for the cylindrical surface and is indicated by means of the direction feature indicator for the conical surface.

The tolerance zone defined by the specification in Fig. 2.28 is limited, in the considered cross-section, by two circles on a conical surface 0,1 mm apart along the surface.

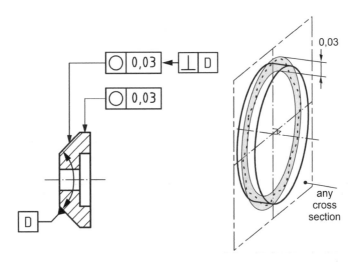

Fig. 2.27   The extracted circumferential line should be contained between two coplanar concentric circles, with a difference in radii of 0,03, for both cylindrical and conical surfaces, in any cross-section of the surfaces perpendicular to an axis. This is the default value for the cylindrical surface and it is indicated by means of the direction feature indicator for a conical surface

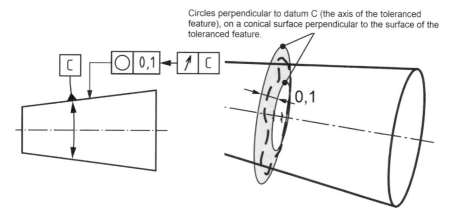

Fig. 2.28 The extracted circumferential line in the conical component on the right should be contained between two circles on the intersecting cone, 0.1 apart. The tolerance zone is perpendicular to the surface of the toleranced feature, as indicated by the direction indicator

The **collection plane** symbol establishes a family of parallel planes, which identifies a closed boundary of contiguous features covered by the "all around" symbol. Indeed, if a geometrical specification is applied to all the features represented by a closed outline, it should be indicated by the "all around" symbol together with a collection plane indicator in order to make the drawing unambiguous (see Fig. 2.29), especially for 3D annotations.

In Fig. 2.30, as the "all around" symbol is used with a UF modifier (Unified Feature), the tolerance applies to a united feature which includes the features that make up the periphery of the workpiece when seen in a plane parallel to datum B, as indicated by the collection plane indicator. Basically, a collection plane identifies a set of single features whose intersection with any plane parallel to the collection plane is a line or a point.

If the specification applies to any cross section or any longitudinal section of a feature, it should be indicated with the "**ACS**" specification modifier for any cross section or with the "**ALS**" specification modifier for any longitudinal section. The two

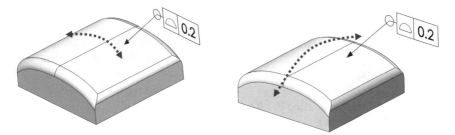

Fig. 2.29 Without any indication of the collection plane, the symbol "all around" symbol is ambiguously applied to all the features represented by a closed outline

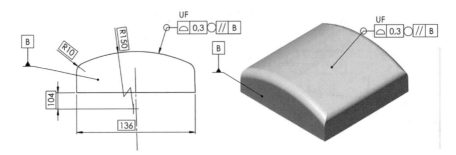

Fig. 2.30 Since the "all around" symbol and the UF modifier are used, the tolerance applies to a united feature consisting of the features that make up the periphery of the workpiece when seen in a plane parallel to datum B, as indicated by the collection plane indicator

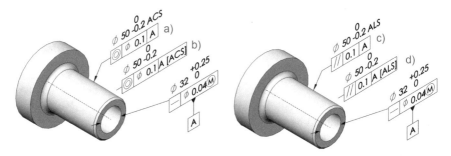

Fig. 2.31 The ACS and ALS symbols can both be placed above the tolerance frame (**a, c**) or after the datum letter symbol in the tolerance frame (**b, d**), with no change in meaning

symbols can be placed above the tolerance frame (see Fig. 2.31a) or after the datum letter symbol in the tolerance frame (see Fig. 2.31b), with no change in meaning.

When the ACS modifier is indicated, the toleranced feature and the datum feature **are defined in the same cross-section**. The intersection plane that defines the toleranced feature is by definition a plane perpendicular to the median feature of the associated feature, as established from the extracted integral surface.

Indicating "ALS" above the tolerance frame (see Fig. 2.31c) or after the datum letter symbol (see Fig. 2.31d), allows the datum feature to be defined as any longitudinal section of the integral feature (concurrently with the toleranced feature). The datum feature is the intersection of the real integral feature used to establish the datum and the section plane. A longitudinal section is defined as a half plane that includes an axis.

Figure 2.32 shows an example of the use of a rank-order specification with ALS and ACS modifiers, according to ISO 14405/1.

Figure 2.33 indicates the use of the **between symbol** adjacent to the tolerance indicator and identifies the two locations where each value applies. In this case, a proportional variation is defined from one value to another, between two specified locations, on the considered feature.

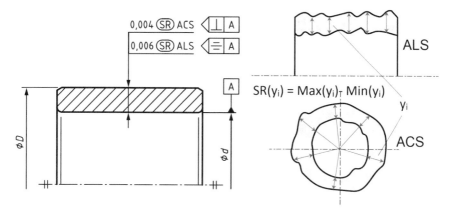

Fig. 2.32 Example of the use of a rank-order specification with ALS and ACS modifiers. For the upper indication, an upper limit (0,004) applies to the range of the two-point size values defined in any cross section. For the lower indication, an upper limit (0,006) applies to the range of the two-point size values defined in any longitudinal section

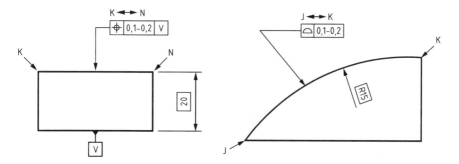

Fig. 2.33 Drawing with indications of the variable width specification using the between symbol

In the ASME Y14.5:2018 standard, the profile tolerance, with a proportional variation between two specified locations on the considered feature, is indicated with the new **From-To** symbol, which is located *below* the tolerance frame. In Fig. 2.34, the tolerance width varies proportionally from 0.1 at S to 0.3 at T.

2.3.3.2   Associated Toleranced Feature Specification Elements

Another important novelty of ISO 1101 is the introduction of new modifiers that allow the tolerance to be applied, not to the *extracted* feature (derived or integral), as the standard default foresees, but to an **associated** feature, utilising various association criteria, with the letters C,G,N, T and X. In fact, in the case of Fig. 2.35, a location error is prescribed to which the real feature should not be subjected, but rather the associated feature, for example, with the maximum envelope criterion (modifier

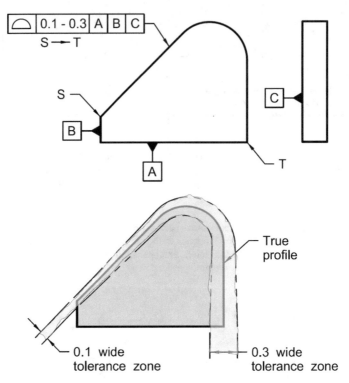

Fig. 2.34 The profile tolerance in the ASME Y14.5:2018 standard, with a proportional variation between two specified locations on the considered feature, is indicated with the new *From-To* symbol located below the tolerance frame. The tolerance width in the figure varies proportionally between 0.1 at S and 0.3 at T

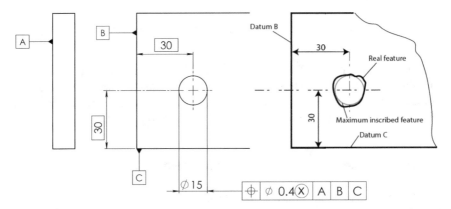

Fig. 2.35 The location error to which the real feature should not be subjected, but an associated feature, for example, a position specification that applies to the associated maximum inscribed feature (modifier X) is instead prescribed here

X), or the associated tangent plane (modifier T, which already existed in the ASME standard). Considering that an associated feature is ideal (and therefore without form errors), this option can only be applied to orientation or location tolerances.

As can be seen in Sect. 2.1, an associated feature is an ideal feature that is established from a real feature through an association operation. By default, the reference feature association is the **minimax** (*Chebyshev*) association without constraints and it can be used for form specifications, but it is also possible, as an alternative, to indicate the reference feature association specification element as in Fig. 2.36. Table 2.4 shows the symbols and the relative meanings of the associated criteria. E, I should be used to indicate the criteria association with the constraint external (E) and internal (I) to the material.

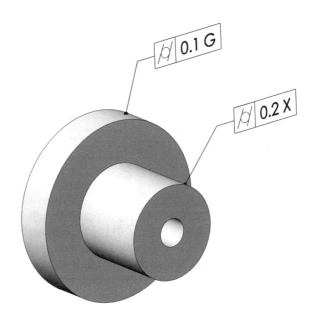

Fig. 2.36 Cylindricity control with indication of the association of the two reference features (maximum inscribed and Gaussian association). The modifier X maximises the size of the reference feature, while maintaining it entirely inside the toleranced feature. G should be used to indicate the least squares (Gaussian) association. It minimises the square of the local deviations of the toleranced feature to the reference feature

Table. 2.4 Associated toleranced feature specification elements. The symbol inside the circle in the second column indicates that the tolerance is applied to the associated element. The symbol in the third column indicates the association criterion and is usually applied to form tolerances

	Associated toleranced feature	Reference feature association
Minimax (Chebyshev)	Ⓒ	C, CE, CI
Least squares (Gaussian)	Ⓖ	G, GE, GI
Minimum circumscribed	Ⓝ	N
Maximum inscribed	Ⓧ	X
Tangent feature	Ⓣ	– –

Fig. 2.37 Roundness
specification with a
minimum circumscribed
reference feature after the
application of a Gaussian
long-wave pass filter with a
cutoff value of 50 UPR

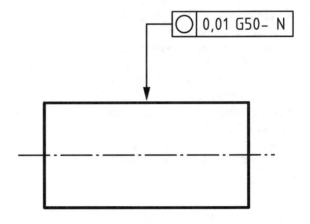

The filter specification is an optional specification element which is indicated by a combination of two specification elements, that is, the type of filter and the nesting index or indices for the filter (Fig. 2.37).

2.3.4 New Symbols and Specification Modifiers of ISO 5458:2018

The new ISO 5458:2018 standard establishes rules that may be considered complementary rules to ISO 1101 to apply to pattern specifications. According to the **independency principle** (ISO 8015:2011, 5.5), a geometrical specification that applies to more than one single feature by default also **applies to those features independently.** The tolerance zones defined by one tolerance indicator or by several tolerance indicators should therefore be considered independently by default.

The new standard has introduced a new definition of internal constraints (location constraint and/or orientation constraints between the individual tolerance zones of the tolerance zone pattern) using the CZ,CZR or SIM modifiers (Table 2.5).

There are two ways of creating a tolerance zone pattern, that is, either by using a single indicator pattern specification with the CZ or CZR modifiers, or by using a multiple indicator pattern specification using SIM modifiers (Fig. 2.38).

Table. 2.5 Symbols and specification modifiers in the ISO 5458:2018 standard

Applied to	Symbol	Description	Constraint
Toleranced feature	UF	United feature	Not applicable
Tolerance zone	SZ	Separate zones	None
	SIM i	Simultaneous requirement No. i	Orientation and location
	CZ	Combined zone	Orientation and location
	CZR	Combined zone rotational only	Orientation constraint only

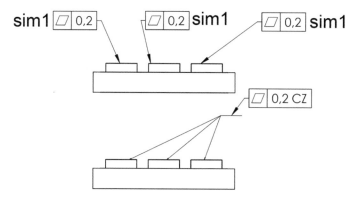

Fig. 2.38 There are two ways of creating a tolerance zone pattern, that is, either by using a single indicator pattern specification with the CZ or CZR modifiers or by using a multiple indicator pattern specification using SIM modifiers

Chapter 3
Dimensioning with Geometrical Tolerances

Abstract The advantages of the geometric specification of a product are here illustrated by converting a traditional 2D coordinate tolerancing drawing using the geometrical tolerance method. The new language of symbols permits the functional requirements of the products to be fully expressed in technical documentation. Moreover the dimensioning of a workpiece according to the geometrical tolerance method can reduce the ambiguity in the indications and in the interpretation of the dimensional and geometrical requirements of the products in order to achieve not only unambiguous communication between the design, production and quality control entities, but also with the clients and suppliers of the outsourced processes.

3.1 Conversion from Traditional 2D Coordinate Tolerancing

Let us now consider the dimensioning of the previous workpiece according to the geometrical tolerance method (Fig. 3.1), utilising symbols and codes that will be dealt with in more detail later on.

In order to apply this method, some simple principles should be taken into account:

(a) Two Cartesian reference systems should be set up (through three orthogonal planes indicated as A, B, C and an axis D in Fig. 3.2) with respect to which all the geometrical components should be located in a univocal manner.

(b) The zone and the tolerance value should be defined for each feature (Fig. 3.3). The real features of the part should fall within the boundary of their theoretical location, that is, the one they have in the ideal component.

(c) The tolerance zones are located and orientated with respect to the reference system: for example, the cylindrical tolerance zone of the central 20 mm hole is oriented perpendicular to datum plane A and located, with Theoretically Exact Dimensions (TED), with reference to datum planes B and C.

(d) Plus or minus tolerancing is only used to define the dimensions of a feature of size[1] (the holes in this case).

---

[1] A **feature of size** (FOS) is a set of two opposite elements associated with a size dimension.

© The Author(s), under exclusive license to Springer Nature Switzerland AG 2021   41
S. Tornincasa, *Technical Drawing for Product Design*, Springer Tracts
in Mechanical Engineering, https://doi.org/10.1007/978-3-030-60854-5_3

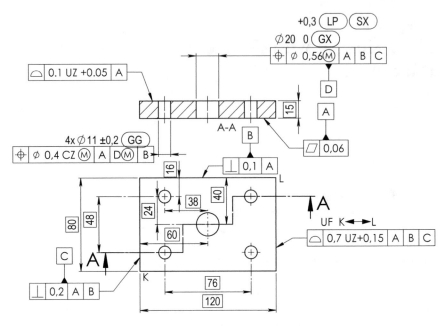

Fig. 3.1 The dimensioning of a workpiece through the geometrical tolerance method: the tolerance zones become clear and univocal and the tolerance zones on profiles are equivalent in amplitude to those obtained by means of traditional dimensioning, but an undesired accumulation of errors is avoided, and the datum sequence is reported immediately and in a simple manner for the inspection

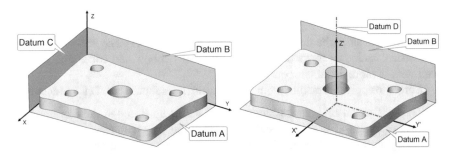

Fig. 3.2 In order to apply the functional dimensioning principles, two Cartesian coordinate systems should be set up, with respect to which all the geometric features of the component will be located

(e) The tolerance zone is increased if a circled M is indicated inside the toler-
     ance indicator of the location tolerance. This means that this tolerance has been
     foreseen to ensure functioning when the hole is under maximum material condi-
     tions (that is, with the minimum diameter) and it is accepted that, if the hole is
     produced with a larger diameter, mating will also be possible, even with a larger
     location error (**Maximum Material Requirement**).

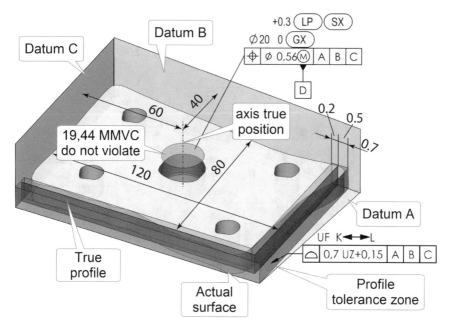

Fig. 3.3 The zone and tolerance value of each feature should be defined. The real features of the part should fall within the boundary of their theoretical location, that is, within the location they have in the ideal workpiece

The feature of size concept
ISO definition: according to ISO 17450/1 a feature of size with a linear size is a geometrical feature that has one or more intrinsic characteristics, only one of which may be considered as a variable parameter. A feature of size can be a sphere, a circle, two straight lines, two parallel opposite planes or a cylinder. A single cylindrical hole or shaft is a linear size feature. Its linear size refers to its diameter. An axis, median plane or centrepoint can be derived from a feature of size.
ASME definition: a regular feature of size is a one cylindrical surface, a spherical surface, a circular element, or a set of two opposite parallel line elements or opposite parallel surfaces associated with a single directly toleranced dimension.
A feature of size:

1. is characterized by opposite points on a surface;
2. an axis, a median plane or a centrepoint may be derived;
3. is associated with a size or a mating dimension.

Which features in Fig. 3.4 are features of size?

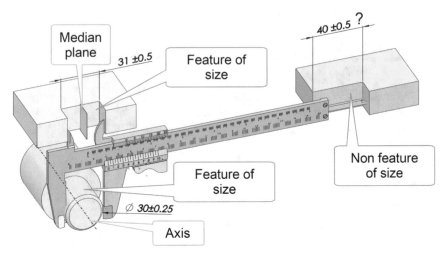

Fig. 3.4 The characteristics of a feature of size

3.2 Advantages of Geometrical Product Specification

Several advantages can be achieved from the application of this type of dimension:

(1) a cross section of a cylindrical tolerance zone circumscribed about a square tolerance zone, like the one shown in Fig. 3.5, has 57% more area than the square zone, in which a point of the extracted median line of the hole may lie. It can in fact be noted in Fig. 3.5 that, for the same maximum amplitude of admissible tolerance, that is, 0.56 (as shown in Fig. 1.6), even those features that fall outside the tolerance range would be acceptable, according to what is indicated in Fig. 1.6, because they have a point of extracted median line located inside the round segments;

(2) the holes are located with respect to the three-plane datum system, and this means that, for the control, the workpiece should first be placed in contact with datum A, and then in contact with datum B and finally fixed with datum C (Fig. 3.6); in this way, the control is univocal and repeatable, even when carried out in different periods of time and by different workers;

(3) no problem of tolerance accumulation arises, as all the location dimensions refer to theoretically exact dimensions;

(4) in certain cases, by applying the **MMR**[2] requirement, some tolerances can be doubled; in fact, the location tolerance of a hole (0.56 mm) becomes 0.86 mm when the hole is produced under the minimum material condition (20.3 mm);

(5) the dimensioning and tolerancing of each feature should be complete and clearly defined, so that the form, orientation, location, and where applicable, the size of each feature on a part are fully defined on a part. It is possible to directly

---

[2]See Sect. 5.1.1.

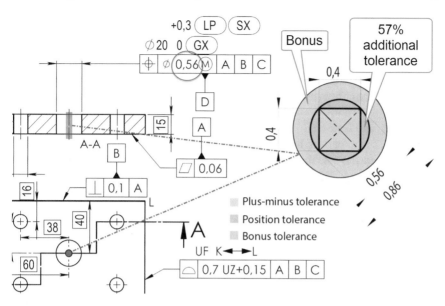

Fig. 3.5 The location tolerance of the extracted median line of the hole defines a cylindrical zone with an increase of 57%, compared to the coordinate dimensioning tolerance shown in Fig. 1.3. A further bonus of 0.3 mm is obtained when the hole is produced under the minimum material conditions (as the tolerance is indicated with the maximum material requirement symbol)

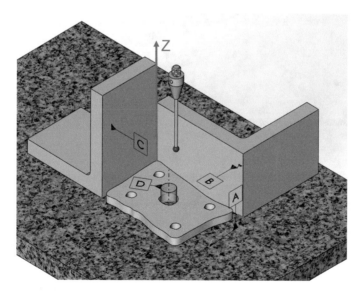

Fig. 3.6 The holes are located with respect to the three-plane datum system. This means that, for the control, the workpiece is first placed with reference to datum A, then moved against datum B and finally fixed with datum C; in this way, the control is univocal and repeatable, even though carried out in different periods and by different workers

locate the dimensioning on a 3D CAD model, as shown in Fig. 3.7, where the theoretically exact dimensions are those of the native geometry of the model.

In short, more reduced and poorer quality functionality and performances of products are obtained with the traditional design (that is, with the coordinate dimensioning system), and the correct assembly of the components is not always guaranteed. Moreover, ambiguous and misunderstanding indications are given for the manufacturing of the parts, and their control is often random.

Table 3.1 summarises the differences between the two dimensioning systems. Companies, in the actual production context, should adopt advanced specification

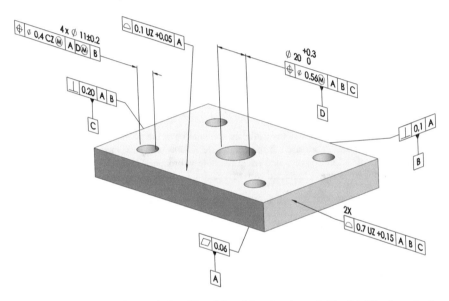

Fig. 3.7 Dimensioning of an equivalent 3D solid model to that shown in Fig. 3.1. The theoretically exact dimensions are those of the native geometry of the model

Table 3.1 Comparison between plus-minus tolerancing and GD&T tolerancing

	Plus-minus tolerancing	GPS tolerancing
Inspection method	• Multiple inspections may produce different results • Good parts may be scrapped	• The datum system communicates the right set up for inspection • Clear instruction for inspection
Tolerance zone	• The tolerance zone is fixed in size • Functional parts may be scrapped	• The use of modifiers allows the tolerance zone to be increased • Disputes over part acceptance are eliminated
Tolerance zone shape	• Square or rectangular tolerance zone for hole locations	• A diameter symbol allows round tolerance zones to be obtained with +57% more tolerance

and geometrical verification methodologies, based on the most recent developments of the international ISO and ASME standards, for their products in order to:

(1) *Fully express the functional requirements of the products* in the technical documentation;
(2) *Reduce ambiguity in the indications and in the interpretation of the dimensional and geometrical requirements* of the products in order to achieve unambiguous communication between the design, production and quality control entities, but also with the clients and suppliers of the outsourced processes.

Ultimately, GPS (or GD&T) **is a symbolic language that can be used to search for, refine and encode the function of each feature of a part in the design phase**, with the objective—through the decoding process—of guaranteeing assembly and functionality, specifying the manufacturing objectives, reducing the production costs and of transforming the control into **a real scientific and reliable process**.

Chapter 4
The GPS and GD&T Language

Abstract This chapter is focused on the main differences between the ISO and ASME standards in the geometrical specification domain of the industrial products, and it starts with details on the historical evolution of the two standards. The main principles of the ISO GPS and ASME GD&T standards, such as the principle of independence and the envelope requirement, are illustrated. Designers are recommended to always indicate the reference standard in the technical drawings of companies, as the interpretation of drawing specifications and the relative inspection may lead to two different results. Finally, the main novelties of the new ASME Y14.5:2018 standard and the new ISO 22081 standard on general tolerances are shown.

4.1 Historical Evolution of the ISO and ASME Standards

The current industrial situation is characterised more and more by a continuous evolution towards increasingly dynamic interaction models between clients and suppliers that put traditional technical communication methodologies under greater pressure.

An always increasing requirement of accuracy in the description and in the interpretation of the functional requirements, and consequently in the drawing up of the technical designs and documents in the mechanical subcontracting sector, which are coherent and complete and able to adequately support the co-design and outsourcing requirements of a production, has been observed.

For this reason, a remarkable effort is under way to develop a coherent and innovative management scheme of geometric tolerances, in order to obtain a better definition of the correlation between the functional requirements, geometrical specifications and relative control procedures, which can be summarised as the Geometrical Product Specification—GPS and Geometric Dimensioning and Tolerancing – GD&T principles, and which, if implemented correctly and coherently, allow the drawbacks of the present methodologies to be overcome and intra and inter-company communication to be revolutionised.

The GPS and GD&T methodologies have evolved according to two basically different approaches (Fig. 4.1). A single standard was developed in the ASME standard, at the end of the 1960's, to define the fundamental regulations for functional

© The Author(s), under exclusive license to Springer Nature Switzerland AG 2021
S. Torincasa, *Technical Drawing for Product Design*, Springer Tracts
in Mechanical Engineering, https://doi.org/10.1007/978-3-030-60854-5_4

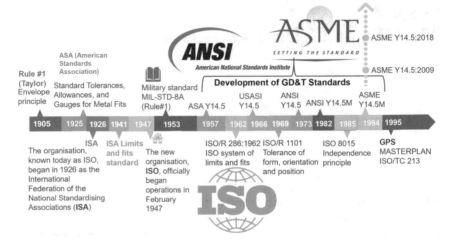

Fig. 4.1 Historical evolution of the ASME and ISO standards on geometrical tolerance

dimensioning. The objective was that of creating a clearly defined and coherent normative system, which gave rise to the ASME Y14.5 standards of 1994 and 2009, and the ASME Y14.5 standard of 2018.

4.1.1   The Birth of Tolerances

The world's first tolerance standard of limits and fits made use of by industry was published in 1902 in the United Kingdom by the Newall Engineering Co.. In 1901, the UK established the world's first standardisation organization, the Engineering Standards Committee (which, in 1931, became the **British Standards Institution**, *BSI*) and which issued its first standard on limits and fits in 1906. This standard involved the use of a standard sized shaft with various sizes of holes to establish the various types of fit. The standard, which was not initially regarded favourably by industry, since it was based on a shaft basis, was revised in 1924 on basis of the holes.

In the United States, five engineering societies and three government agencies founded the **American Engineering Standards Committee** (*AESC*) in 1918, which then became the **American Standards Association** (*ASA*) in 1928 and which was reorganised as the **United States of America Standards Institute** (*USASI*) in 1966. Finally, the USASI became the **American National Standards Institute** (*ANSI*) in 1969. The first standard on limits and fits was the *American Tentative Standard Tolerances, Allowances, and Gages for Metal Fits, B4a-1925* which contained tables that listed eight series of fits between holes and shafts, each fit being specified by the limits of size for each of the two mating parts.

The *International Federation of the National Standardizing Associations* (or *International Standards Association*, **ISA**), was established in 1926, and one of the earliest projects was the development of an international system of limits and fits, which was published in 1941 as "*ISA Tolerance System*" with all the data in metric units. Several European countries adopted this system as the basis of their national standards.

After the conclusion of World War II, a new organisation, the *United Nations Standards Coordinating Committee* (UNSCC), was established by the United States, Great Britain and Canada to extend the benefits of standardisation to the work of reconstruction. In October 1946, ISA and UNSCC delegates from 25 countries met in London and agreed to join forces to create the new **International Organisation for Standardisation** (*ISO*). The new organisation, ISO, officially began operations in February 1947 with its first office in Geneva, Switzerland.

In the same period, in the United States the geometrical tolerances language started as a military standard, known as US *Army 30–1–7*, dated April 15th, 1946, and was then updated as *Mil-Std-8* in 1949.

The **American Society of Mechanical Engineers** (*ASME*), a non-profit organisation which was founded in 1880, is one of the oldest standards-developing organisations in America. In 1957, ASME published the first dimensioning and tolerancing standard, that is, **Y14.5–1957** for ASA. Subsequent revisions of the Y14.5 standard have been published as USASI (**Y14.5–1966**) and ANSI (**Y14.5–1973** and **Y14.5–1982**).

The revised ASME **Y14.5 M-1994** ("M" because metric units were included) was approved as an ASME standard and, after fifteen years, was followed by **ASME Y14.5–2009**. The current version is **ASME Y14.5–2018,** and it was released in February 2019. The objectives of each version have been to correct any inconsistencies in the previous edition through the work committees made up of volunteers from academia and industry.

Since its foundation, ISO has developed many standards in the field of geometric tolerances such as the ISO 8015 standard in 1985 (*principle of independence*) and the ISO 1101 standard in 1969 (*form and position tolerances*).

In the ISO ambient of the 1990's, starting from the consideration that 50% of the standards necessary for GD&T dimensioning were not available or were even in contradiction with the other existing standards, the international technical-scientific community was stimulated into searching for a new, more general and richer language, constructed on the basis of rigorous mathematical assumptions, that is, the aforementioned GPS.

In 1993, ISO set up the *Joint Harmonisation Group* (ISO/TC 3-10-57/JHG), in which the pre-existing technical committees, that is, ISO/TC3 (Surface Texture), ISO/TC10 (Dimensioning and Tolerancing) and ISO/TC57 (Measurement) were joined together in order to prepare a new standard. Between 1993 and 1996, the Joint Harmonisation Group developed the plant philosophy of the new language and, in 1995, the ISO/TR 14638:1995 "**Masterplan**" document was issued, which contains, among others, the proposal of a new paradigm for the classification of the existing and future standards, starting from the consideration that 50% of the standards necessary for GD&T dimensioning are not available or that they contradict other existing standards.

In 1996, the Joint Harmonisation Group was disbanded in Paris and the ISO/TC 213 technical Committee was set up. Some American representatives of the ASME Y14.5 committee took part in ISO committee meetings until 1999, after which they stopped attending because of controversies connected to the definition of certain geometrical concepts in the two normative systems, which adopt very different approaches, in spite of some apparent analogies.

4.2   The GPS Matrix Model

The objective of the new language was that of expressing and transmitting, in a rigorous formal manner, all the functional requirements of the products, in order to guarantee functionality, reliability, verifiability and interchangeability. GPS is considered as a shared language between worlds that are often separate, that is, the design, production and the control worlds and, for the first time in the history of standards, it compares designers with metrologists. The ISO/TR 14638:1995 Masterplan document, which was issued in 1995 as a summary of the work of the Joint Harmonisation Group, was drawn up to outline the guidelines for ISO/TC 213. In fact, the Masterplan, which was later approved as the Global GPS Standard, ratifies the new paradigm for the classification of GPS standards. The entire system is known as the "**GPS Matrix Model**", and it takes on the role of a "container" of the GPS standards by acting as a matrix in which the lines refer to the geometrical properties of the product (for example, the form or location), while the columns represent the links or, in other words, a specific application ambit of the standards in the context of the development cycle of a product, from its conception to its final control (Fig. 4.2).

A table is formed by crossing these dimensions: the GPS matrix, in which each standard is characterised by 2 coordinates (the properties and the production process step). A standard can often refer to more than one property, or to different steps of the production process; it can therefore take up an area of the matrix and not just a single cell. Each standard of the system includes a final attachment with the exact location of the standard in the matrix, which is indicated with a filled dot (Fig. 4.3).

In the new Masterplan outlined in ISO 14638 of 2015, the GPS standards are classified as:

1. **Fundamental**, that is, standards which define the rules and principles that apply to all categories and which occupy all the segments of the matrix (ISO 8015, ISO 14638).
2. **General**, that is, ISO GPS standards that apply to one or more geometrical property categories, and to one or more chain links, but which are fundamental (ISO 1101, ISO 5459).

Table 1 — ISO GPS Standards matrix model

	Chain links						
	A	B	C	D	E	F	G
	Symbols and indications	Feature requirements	Feature properties	Conformance and non-conformance	Measurement	Measurement equipment	Calibrations
Size	Geometrical features				Comparison	Verification	
Distance							
Form							
Orientation							
Location							
Run-out	Specification						
Profile sur-face texture							
Areal surface texture							
Surface imperfections							

Fig. 4.2 GPS matrix model of the ISO 14638:2015 standard where the lines refer to a specific geometrical feature of the product, while the columns represent the application environments of the standards in the product development cycle context

	Chain links						
	A	B	C	D	E	F	G
	Symbols and indications	Feature requirements	Feature properties	Conformance and non-conformance	Measurement	Measurement equipment	Calibration
Size							
Distance							
Form	●	●	●				
Orientation	●	●	●				
Location	●	●	●				
Run-out	●	●	●				
Profile surface texture							
Areal surface texture							
Surface imperfections							

Fig. 4.3 Each standard is characterised by 2 coordinates in the matrix, but can often refer to more than one property, or different steps of the production process, and may thus occupy an area of the matrix rather than just a single compartment, as is the case of ISO 1101 of 2017

3. **Complementary**, that is, ISO GPS standards that refer to specific manufacturing processes (for example, turning) or to specific machine elements (for example, bolts).

The applicability hierarchy was conceived in such a way that the highest standards (general) also apply for the lowest (specific) ones. For example, it is not necessary to specify the reference temperature in ISO 1101 (general, geometrical tolerance) as UNI ISO 1 (fundamental) is valid.

4.3   The Fundamental ISO 8015 Standard

In the 1980's, the international practices relative to the application of tolerances had, for some time, pointed out the necessity of defining the relationship between dimensional and geometrical tolerances, which had led to an ISO standard in 1985, then transposed by UNI as a national standard, that is, UNI ISO 8015 in 1989, and updated in 2011.

It should be noted that the UNI 7226 standard in the 1973 text defined the relationship between geometrical and dimensional tolerances as: "*When only dimensional tolerances are foreseen, all the deviations of form are limited by the dimensional tolerances; in other words, the real surfaces of the objects can deviate from the functional geometrical form as long as they remain within the dimensional tolerances. If the form errors fall within such limits, the form tolerance shall be indicated*".

This concept creates problems if applied to tolerances that can be associated, such as perpendicularity or parallelism, and the section regarding this dependency principle between form and size was eliminated in the revised ISO 1101 standard and consequently in UNI 7226/1.

The "**Independency principle**" was introduced as a substitute for the aforementioned ISO 8015 standard of 2011, as the fundamental principle for the assignment of tolerances, according to which "*Every dimensional or geometrical prescription specified on a design must be respected in itself in an independent way, except when a particular relation is specified*".

Therefore, when no specific indications are given, geometrical tolerances should be applied without taking into consideration the sizes of the element, and its prescriptions (dimensional and geometrical) should be treated as requirements that are independent of each other. In this way, deviations of form are no longer limited by the dimensional tolerances.

In this context, let us consider the shaft shown in Fig. 4.4; apart from the dimensional tolerance on the 50 mm diameter, a roundness tolerance appears in a tolerance indicator. As the geometrical tolerances are no longer constrained to the dimensional ones, the shaft may have a lobed form inside the round tolerance, where all the sections are under a condition in which all the sections have the maximum dimension permitted for such a dimensional tolerance. In the worst case, the virtual size (mating size) amounts to 50.02 mm (50 + 0.02).

According to ISO 8015 of 1989, the drawing for which the independency principle is applied should be appropriately identified in order to avoid confusion with the previous designs. Moreover, they should report the following indication:

TOLERANCING ISO 8015

The ISO 8015 standard of 2011 reaffirms the independency principle, and even clarifies that, by default, each GPS prescription pertaining to a feature or to a relationship between a feature and a component should be considered completely independent of other specifications, except in the case in which modifiers such as M and E are used.

Fig. 4.4 Interpretation of a drawing according to ISO 8015: a lobed form may appear within a round tolerance in a shaft where all the sections are under a condition in which all the sections have the maximum dimension permitted by the dimensional tolerance. In the worst case, the virtual size (mating size) amounts to 50.02 mm (50 + 0.02). It should be noted that the straightness tolerance (0.06) is greater than the dimensional one (0.039)

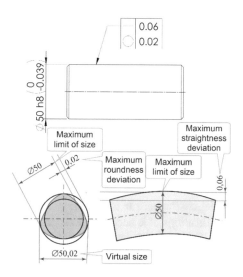

This means that, in order to indicate the application of the independency principle, it is no longer necessary to indicate the words "Tolerancing ISO 8015" in correspondence to the title block of a drawing. As many GPS symbols are identical to the GD&T symbols of the ASME standards, it would be convenient, to avoid confusion, to write the following in the title block:

ISO 8015 or *TOLERANCING ISO 8015*

The "*ISO default GPS specification*" concept, which is defined by the ISO standard, and the "*altered default GPS specification*" concept, which is a GPS specification that is defined by other standards, are also introduced.

In the latter case, the standards recommend indicating the use of a non-GPS standard in the drawing, in order to make the interpretation clear and unambiguous.

The indication should therefore foresee (Fig. 4.5):

- the word "Tolerancing" or "Tolerancing ISO 8015";

	FINITURA: Ra 3.2		POLITECNICO DI TORINO	Oggetto: Quotatura		B
				Disegnatore	Codice	
				Sandro Rossi	234567	
	DATA		Tav. N. 1	Descrizione:		
DISEG.	10/04/2018		Lavorazione			
File	28/06/2015 19:49:24		Grezzo		Montante	
APPR.						A
FABB.						
Qual.			MATERIALE:	Nome file:		A3
			Lega 1060	453456.1.2.4		
Dimensioni: ASME Y14.5:2009		PESO (kg): 141,90		SCALA:1:1	FOGLIO 1 DI 1	

Tolerancing ISO 8015 (AD) ASME Y14.5:2018

Fig. 4.5 Indication of a specific GPS that is different from the ISO one

- the **AD** symbol that indicates "altered default";
- indication of the *non-ISO GPS standard* that is considered, including the date of emission.

For example, in the case in which the ASME Y14.5 standard of 2009 is used, the wording should be:
Tolerancing ISO 8015 ASME (AD) Y14.5:2009

4.3.1   The Effect of ISO 8015 on Technical Documentation

In the case where it is necessary to indicate the previous versions of such GPS specifications, it is necessary to specify them unambiguously, i.e. *"TOLERANCING ISO 8015:1985"* instead of *"TOLERANCING ISO 8015"*. This will be necessary in situations where an old drawing has been revised and this drawing is now valid with the respective modifications, although not updated in the subsequent GPS standards.

In fact, the second edition of the ISO 8015 standard of 2011 represents a real revolution of the concepts expressed in the previous 1985 edition and defines a real temporal barrier between the drawings conceived before and after 2011 (Fig. 4.6).

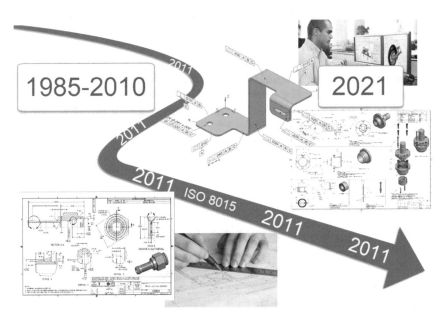

Fig. 4.6 The second edition of the ISO 8015 standard of 2011 represents a revolution of the concepts expressed in the previous 1985 edition and defines a real temporal barrier between the drawings conceived before and after 2011

Principle 5.1 of the standard (the **Invocation principle**) stipulates that "*Once a portion of the ISO GPS system is invoked in mechanical engineering product documentation, the entire ISO GPS system is invoked, unless otherwise indicated on the documentation*", that is, a single reference of the GPS language (such as, for example, a dimension indicated with 30H7) **invokes the entire GPS system** (unless otherwise indicated).

When the GPS system is invoked, there is a series of consequences, including:

- All the principles of ISO 8015 are applied;
- The normal reference temperature is fixed by ISO 1 as 20 °C.

Figure 4.7 shows a drawing drawn up in 2001, for which ISO 8015, not being explicitly cited, the envelop requirement holds, that is, the perfect (geometrically ideal) envelope with the maximum material size of the dimensional tolerance interval (which also represents the worst mating conditions) [1].

The same drawing, released in 2012 (Fig. 4.8), should be interpreted in a rigorously different manner. The ISO 8015 standard *is a default standard*, and there is therefore no need to indicate it.

The drawing invokes the two general ISO 1302 and ISO 13715 standards and consequently the entire GPS system holds.

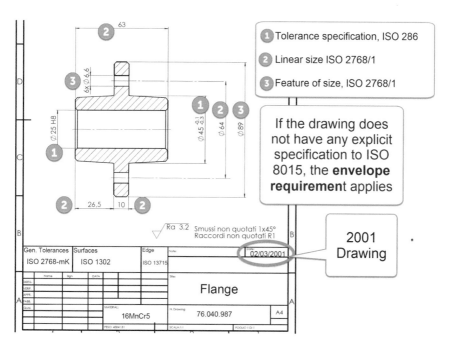

Fig. 4.7 Drawing prepared in 2001, in which ISO 8015 of 1985, not being explicitly cited, the envelope principle holds, that is, the perfect (geometrically ideal) envelope with the maximum material size of the dimensional tolerance interval, which also represents the worst mating condition

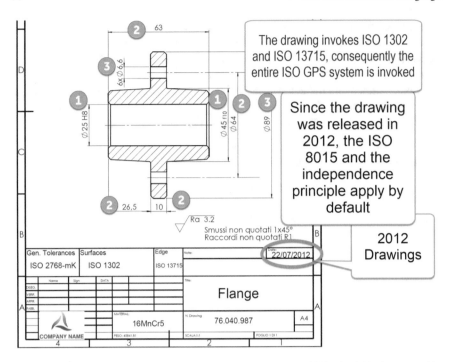

Fig. 4.8 The same drawing as before, but this time conceived in 2012, should be interpreted in a rigorously different manner: ISO 8015 is a default standard, and it is therefore not necessary to indicate it. The design invokes the two general standards, that is, ISO 1302 and ISO 13715, and consequently the entire GPS system holds

1. The dimensions indicated with number 1 recall ISO 286-1:2010.
2. The other dimensions recall 14405–1:2016, and the control therefore involves the distance between two points (local size, default).
3. The independency principle is applied, that is, the worst mating dimension is not that of the maximum material.

The 25H8 dimension in Fig. 4.9 has a tolerance size of 33 μm, while its maximum material limit (minimum size) amounts to 25 mm. When referring to 2768-K, the deviation of the cylinder from roundness may be equal to the tolerance of size, that is, 33 μm. In the worst case, the virtual size (mating size) amounts to 24.967 mm (25–0.033). What advice can be given to designers to avoid this problem? In order to avoid the application of the independency principle, the envelope principle can be suggested with the general indication: **Linear sizes ISO 14405 Ⓔ**

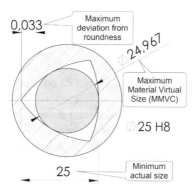

Fig. 4.9 Consequences of the independency principle, which could lead to judicial disputes with the suppliers: the Ø25 H8 dimension has a tolerance of 33 μm and a maximum material dimension of 25 mm. According to ISO 2768-K, the roundness tolerance should be equal to the tolerance, that is, to 33 μm. The mating dimension should be equal to 24.967 mm (25–0.033 mm)

4.3.2 The Main Concepts Defined in ISO 8015:2011

Other important concepts that are defined in the ISO 8015 standard of 2011 are:

1. Fundamental assumptions for the reading of specifications on drawings (Sect. 4):

    (a) *Functional limits* (Sect. 4.2), based on experimental and/or theoretical investigations, and which are known with no uncertainty. The choice of the designer, pertaining to defining the limits within which the overall functionality is reached, can never be justified.
    (b) *Tolerance limits* (Sect. 4.3), which are identical to the functional limits, which means the designer has the responsibility of indicating the functional limits on the drawing through the tolerance limits.
    (c) *Workpiece functional level* (Sect. 4.4), which means that the workpiece should function 100% within the tolerance limits and 0% outside the tolerance limits.

2. **Definitive drawing principle** (Sect. 5.3): if not indicated, the specifications are applied to the final stage of the product, as designed, but intermediate production stages can also be indicated.
3. **Feature principle** (Sect. 5.4): a workpiece should be considered to be made up of several features limited by natural boundaries. Each and every GPS specification of a feature or relationship between features applies by default to the entire feature or features; each GPS specification applies only to one feature or one relation between features. Indications which specify that a requirement applies to more than one feature, e.g. when the *CZ* (combined tolerance zone) indication is used, are available.
4. **Independency principle** (Sect. 5.5): each and every GPS specification of a feature or relation between features should by default be fulfilled independently

of any other specifications, except when a special indication is specified (e.g. Ⓜ,Ⓛ,Ⓔ, CZ modifiers)

5. **Default principle** (Sect. 5.7): each specific GPS invokes a series of rules that were pre-defined in other standards (default). A rule (the most common) is univocally pre-defined in the GPS system, and it can be omitted from the drawing in order to prevent overloading it. Pre-defined rules can be left out from the drawing by introducing symbols or others (modifiers) defined in the GPS system or even personalised within a company. For example, a 30H6 dimensioned hole (ISO 286–2) should defer to the definition of the tolerance of a diameter of a hole of between 30 and 30.016. As nothing else is specified, the diameter measured as the distance between two points is considered valid (ISO 14405–1).

6. **Reference condition principle** (Sect. 5.8): all GPS specifications apply by default for a standard reference temperature of 20 °C, as defined in ISO 1, and the workpiece should be free of contaminants.

7. **Rigid workpiece principle** (Sect. 5.9): each component is considered as an infinite rigid body that is not affected by any external force. A workpiece should by default be considered as having infinite stiffness, and all GPS specifications apply in the free state, and as undeformed by any external forces, including the force of gravity. A rigid body is a component that does not deform or bend by a quantity that prevents it from functioning under the effects of forces and/or the mounting constraints. Mention is also made of free state, that is, all the dimensions and the specified tolerances are applied without the action of any force other than gravity. When the workpiece is flexible, this should be indicated on the design (ISO 10579-NR).

8. **Duality principle** (Sect. 5.10): the ISO GPS standards express ***duality between the specification and the verification***. Everything that is done in the specification process is reflected in the actual measurement process. The procedure that allows the description of the product to be made in the design and control phases is the same as it utilises the same type of operators.

9. **General specification principle** (Sect. 5.12). General tolerances can be indicated on the design by means of references to the specific standard (e.g. ISO 2768/1) in the title block. These references are applied to each characteristic of each feature or relationship, unless a specific tolerance, which prevails, is indicated. More than one general tolerance can be present, on condition that it is clearly indicated to what each one refers. In the case of conflict, the most permissive prevails.

4.4   General Geometric Tolerances

All the features of a product always have a dimension and a geometrical form; since the functional requirements state that the dimensional deviations and the geometrical deviations should be defined and limited, the drawings should be completed with all the necessary tolerances. However, considering that the geometric tolerances are no longer limited by size tolerances, as a result of the introduction of the independency

Table 4.1 General tolerances according to ISO 2768/2

Tolerance class	Circular run-out tolerances
H	**0,1**
K	**0,2**
L	**0,5**

principle, they should all be indicated on a drawing, which in turn would then be overloaded with too many indications and therefore not so easy to interpret.

It should be recalled that ISO 2768/1 indicates the general dimensional, linear and angular tolerances, which are grouped together in the **f, m, c** and **v** precision classes.

The ISO 2768/2 standard of 1989 had the purpose of simplifying drawing indications and of specifying general geometrical tolerances in order to control those features on a drawing that do not have any respective individual indications. It divides general geometrical tolerances into three tolerance classes (**H, K** and **L**, in decreasing precision), thereby allowing the geometrical specifications and the reading of the drawings to be simplified, and at the same time facilitating the choice of the tolerances.

Table 4.1 shows the values of the general tolerances of a circular run-out, which can also be used for roundness. As no specific values are foreseen in the standard for the roundness tolerance, it is made equal, in numerical value, to the dimensional tolerance on the corresponding diameter (Fig. 4.10), but with the constraint of not exceeding the value of the circular radial run-out, which is instead specified in the standard.

The ISO 2768/2 standard was replaced by the ISO 22081 standard, since it is possible that the general geometrical tolerance indication can make it difficult to interpret the errors and the relative control phase:

1. The standard only prescribes the tolerances values of straightness, flatness, perpendicularity, symmetry and circular run-out.
2. It should be pointed out that the general tolerances do not control the cylindricity, angularity, coaxiality, profile, positional tolerance or total run-out tolerances.
3. Let us consider the same figure that is reported in the ISO 2768/2 standard (Fig. 4.11). A great stretch of imagination is necessary to ascertain that, interpreting the general quality tolerances H, it is possible to define a datum C (hole axis) with respect to which the perpendicularity errors of the two milled faces are specified.
4. The general tolerances indicate orientation errors, but without establishing the datum.
5. Another error is that of the definition of the general angular error (by using the ISO 2768/1 standard, as shown for the workpiece indicated in Fig. 4.12). Should the workpiece be placed on its largest or smallest side in order to perform a control? The tolerance zone in fact becomes larger and larger as the distance from the datum feature increases.

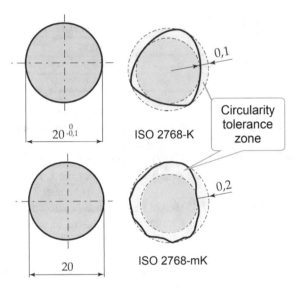

Fig. 4.10 The permitted deviation of the diameter is indicated directly on the drawing; the general circularity tolerance is equal to the numerical value of the diameter tolerance but it is subject to the constraint of not exceeding the value of the circular radial run-out, which is instead specified in the standard. The general tolerances indicated with the ISO 2768-mK specification apply in the figure below. The deviations permitted for a 20 mm diameter are ±0.2 mm. These deviations lead to a numerical value of 0.4 mm, which is greater than the value of 0.2 mm given in Table 4.1; the value of 0.2 mm therefore applies for the circularity tolerance

4.4.1   The New ISO 22081 Standard

The new standard outlines rules pertaining to the definition and interpretation of general specifications defined according to ISO 8015 (general tolerancing) which are applicable to the whole workpiece.

The general geometrical and dimensional specifications can be applied to integral surfaces (not integral lines), according to the following rules:

(a) The general dimensional specifications are applied to features of size, that is linear size (according to ISO 14405-1) and angular size (according to ISO 14405–3).

(b) The only general geometrical specification is the surface profile, which is applied to integral features. In this case, **a datum system should be specified**

The standard emphasises that it is the responsibility of the designer to ensure the **complete** and **unambiguous** definition of functional requirements. Moreover, the geometrical features that influence the functions must be defined appropriately and the **entire workpiece should be completely specified**.

General dimensional specifications are applied to a feature of size that has been identified on a drawing by means of a linear or an angular size *which has no individual*

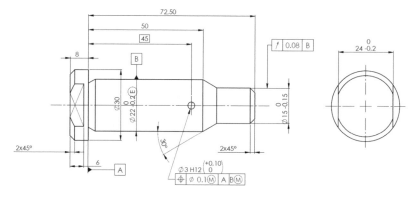

General tolerances ISO 2768-mH

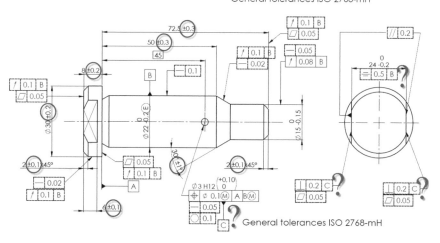

General tolerances ISO 2768-mH

Fig. 4.11   If the figure reported in the ISO 2768/2 standard is taken into consideration, it is possible to demonstrate that the indications of the general tolerances make it difficult to interpret the errors and the relative control phase (the blue circles and the broken-lined frames at the bottom of the figure show the interpretation of the general tolerances). A great effort of imagination is in fact necessary to establish that, interpreting the general tolerances of quality H, it is possible to define a reference C (hole axis) with respect to which the perpendicularity errors of the two milled faces are specified

tolerance and which is not a theoretically exact dimension or an auxiliary dimension (Fig. 4.13).

The **general geometrical specification** applies to each integral feature on the workpiece, with the following exceptions, see Fig. 4.14:

- it does not apply to any feature of size
- it does not apply to any feature with an individual indicated geometrical specification;
- it does not apply to any datum feature referenced in the datum section of the general geometrical specification.

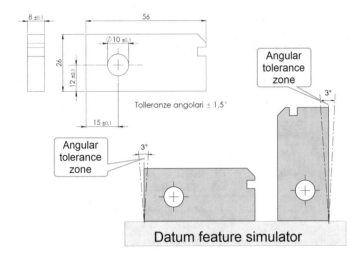

Fig. 4.12 Another error is that of the definition of the general angular error, as shown in the plate above. In order to conduct the control, should the workpiece be placed on its larger or smaller side? The tolerance zone in fact becomes larger and larger as the distance from the datum feature increases

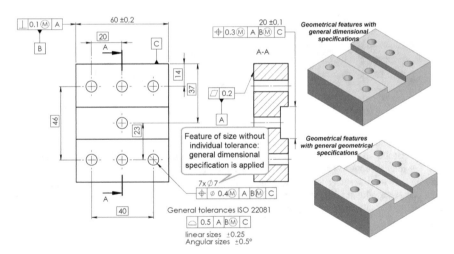

Fig. 4.13 An example of the application of general dimensional and geometrical specifications on integral features

The permissible error for the indication of general dimensional tolerances in technical product documentation can be defined (near the title block) as both a fixed value and as a variable value.

The variable tolerance value should be defined directly by the designer, who should refer directly to a table or a document associated with the drawings (Fig. 4.15).

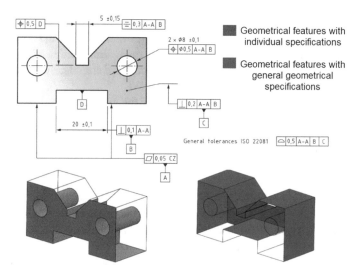

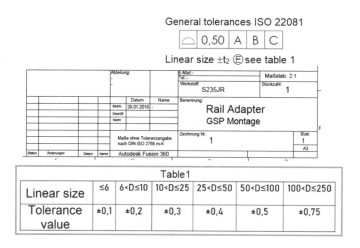

Fig. 4.14 An example of the application of general geometrical specifications to integral features

General tolerances ISO 22081

| ◯ | 0,50 | A | B | C |

Linear size ±t₂ Ⓔ see table 1

Table 1

Linear size	≤6	6<D≤10	10<D≤25	25<D≤50	50<D≤100	100<D≤250
Tolerance value	±0,1	±0,2	±0,3	±0,4	±0,5	±0,75

Fig. 4.15 An example of the indications near the title block, with reference to a table on a drawing

4.5 The New ASME Y14.5:2018 Standard

According to many designers, the ASME Y14.5 standard is the clearest and most coherent standard for the technical documentation of a product, because it uses the GD&T language to communicate the intent of a designer in a complete and unambiguous way, thereby allowing components to be obtained with shapes and dimensions that can guarantee the desired assembly, functions, quality and interchangeability.

The new ASME Y14.5:2018 standard is more readable, more detailed and clearer than the previous version (12 sections, 328 pages compared to the 214 pages of the previous ASME Y14.5:2009 standard), and many ambiguities of the previous version have been eliminated, the layout of the standard completely revised and all the sections have been renumbered.

The standard has also been revised in view of the great changes that have taken place in the twenty-first century, with the widespread use of *Computer-Aided Design* (CAD) and the transition of the industry towards **Model Based Definition** (MBD) techniques with 3D annotations. From this perspective, the GD&T specifications that were previously introduced on many illustrations have been added to the 3D views of the model.

The 14 classical geometric control symbols have been reduced to 12, since **the concentricity and symmetry symbols have been eliminated** and replaced by the position symbol. However, this deletion of the symbols does not leave industry without a means of controlling coaxial or symmetrical features, but it does eliminate the confusion that surrounded these symbols because they were often used incorrectly or were misinterpreted.

The sections dedicated to orientation, form and profile tolerances have been completely restructured to obtain a better readability. In particular, the **FROM-TO** symbol has been added to indicate a specific transition in a non-uniform profile tolerance (Fig. 4.16). Another important change concerns the addition of a new modifier, called "**dynamic profile**". This is a small triangular symbol that can be inserted inside the feature control frame after the tolerance value. The function of the dynamic profile is to allow the form to be controlled, independently of size. Finally, run-out tolerance **may now be used in an assembly** and applied to a tangent plane for one or more coplanar feature faces that are perpendicular to a rotation axis.

Fig. 4.16 The new ASME Y14.5:2018 standard with the new symbols

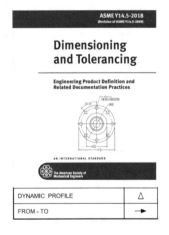

4.5.1 The Envelope Requirement or Rule#1 of the ASME Standard

The ASME Y14.5 standard has codified the **envelope requirement** as Rule#1, whereby *"No element of a feature shall extend beyond the Maximum Material Condition (MMC) boundary of perfect form"*. The envelope requirement may also be applied with the ISO standards, by means of a circled E symbol, placed next to the tolerance dimension. According to this rule (which is also known as *Taylor's rule*), it is stated that where only a tolerance of size is specified, *the limits of size of an individual feature of size prescribe the extent to which variations in its geometric form, as well as its size, are allowed*. (Fig. 4.17).

The form tolerance increases as the actual size of the feature departs from MMC towards the *Least Material Condition* (LMC). There is no perfect form boundary requirement for the LMC. When inspecting a feature of size that is controlled by Rule #1, both its size and form need to be verified. If not otherwise specified, the Least Material Condition (LMC) is verified by making a two-point check at various points along the cross section, while the Maximum Material Condition (MMC) is verified by checking that the whole feature falls within a maximum material envelope of perfect form.

According to ASME Y14.5:2018 (Sect. 3.57), the actual local size is *the actual value of any individual distance at any cross section of a feature of size*. The ASME Y14.5.1M:1994 standard (*Mathematical Definitions of Dimensioning and Tolerancing Principles*) provides a further clarification about the local size requirements of Rule #1. The actual local size is defined by continuously expanding and contracting spheres, whose centres are situated on a derived median line (Fig. 4.18).

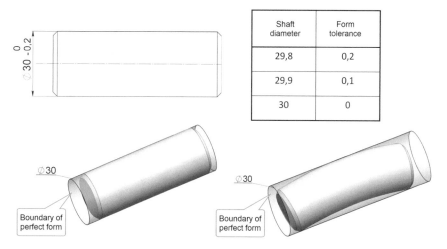

Shaft diameter	Form tolerance
29,8	0,2
29,9	0,1
30	0

Fig. 4.17 Interpretation of the drawings according to Rule#1 of the ASME Y14.5 standard; the limits in variation of the dimensions and form of a shaft allowed by the envelope principle are visible

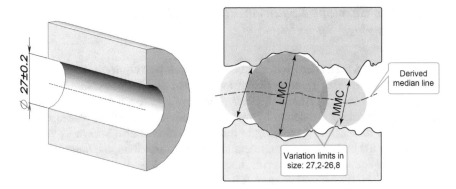

Fig. 4.18 The size of the sphere that may be swept along the derived median line without intersecting the feature boundary determines the actual values of the maximum and least material sizes

Figures 4.19 and 4.20 show the respective control procedures of a hole and a shaft according to Taylor's principle. If the feature fits inside a gauge which has the same size as the maximum material size, the feature conforms to that tolerance limit. At the same time, if the workpiece is at its maximum material condition (MMC), then it cannot undergo any form variation (i. e. out-of-roundness or straightness of its axis), or it will not fit within the boundary.

Figure 4.21 shows the application of the envelope requirement in the ISO standard, where the use of the symbol E within a circle is necessary. Therefore, when all the

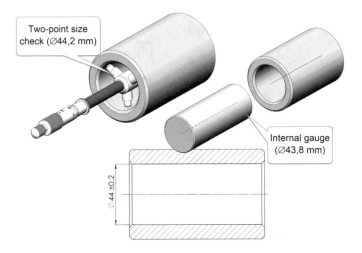

Fig. 4.19 Verification procedure of a hole according to the envelope principle. The minimum material condition is controlled by using an internal gauge (measured between two opposite points), while the maximum material condition is checked by means of a pin with the MMC dimensions

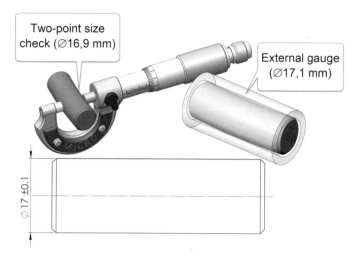

Fig. 4.20 Verification procedure of a shaft according to the envelope principle or ASME Rule#1. The minimum material condition is controlled by using an external gauge (the measurement between two opposite points), while the maximum material condition is checked by means of an envelope of perfect form with MMC dimensions

Fig. 4.21 The envelope requirement in the ISO standard is indicated by means of a circled E, which is placed next to the tolerance dimension; the hole has a perfect form when all the local diameters are under the maximum material conditions, that is, 18 mm

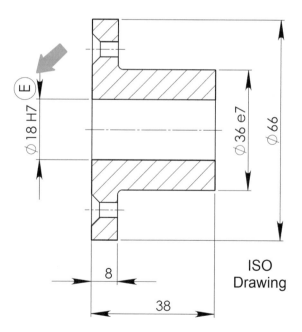

Fig. 4.22 In order to apply
the independency principle
to ASME drawings, it is
necessary to insert the
independence symbol
(circled I) next to the
dimension

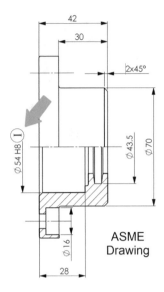

local diameters (18 mm) of the hole are under the maximum material condition, the
form is perfect.

This type of dependence between geometrical and dimensional tolerances ultimately implies that the actual local dimensions of the considered feature vary, within
the assigned dimensional tolerance field, in order to compensate, with reference to
the maximum material dimensions, for any foreseeable form deviations. Such control
standards, which are appropriate in the case of mating, may be restrictive for all the
other geometrical features, and may make it necessary, in the latter case, to furnish
an indication of exception (the ASME standards have introduced the Ⓘ symbol, see
Fig. 4.22), with the consequence of a source of ambiguity being created as it is
not possible to ascertain whether the absence of such an indication depends on the
choices of the designer or rather on an oversight within such a complex technical
document.

Apart from this problem, the verification of the envelope principle, which requires
the use of functional gauges[1] or controls carried out by means of measurement
machines that have been set up and evaluated in an adequate way, is not an easy task.

Table 4.2 shows a summary of the main differences between the two principles. It
should be pointed out that, in the case of ISO GPS standards, considering that, with
the introduction of the independency principle, the form errors are no longer limited
by dimensional tolerances, the general geometric tolerances have been introduced to
simplify the indications on a drawing.

The ASME Y14.5 standard makes the general tolerance indications redundant,
because the envelope principle, which allows the form errors to be limited, is utilised
by default. Figure 4.23 shows the potentiality of such a specification, which allows

---

[1] *Gage* in ASME.

Table 4.2 Comparison of the fundamental ISO and ASME principles

	Advantages	Disadvantages
ISO 8015: Principle of independency	Complete independence between form and size	Requires the use of general geometric tolerances
	The envelope requirement only applies where it is needed	Difficulty of specifying the functional mating
Rule #1: Envelope Requirement	Clearer and more univocal drawings	Inspection with functional gauges
	Simple and easy dimensioning	Restrictive for non-functional geometrical features

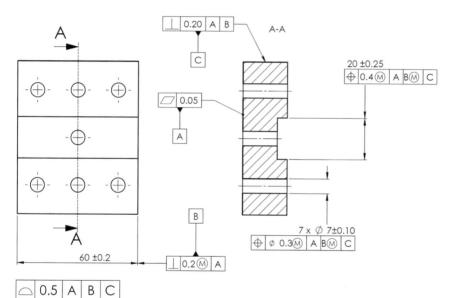

This drawing should be interpreted in accordance with ASME Y14.5-2018
Dimensions not shown on drawing are controlled by means of a 3D CAD file

Fig. 4.23 The ASME Y14.5 standard makes the indications on the general tolerances redundant, as it adopts the envelope principle, which allows the form errors to be limited by default Rule #1

both the dimensional and geometrical general tolerances to be eliminated, thus making the control phase univocal and coherent. Moreover, the tolerance of a profile locates and orients all the surfaces of the workpiece (except the datum features, which have other specific controls [2], see Fig. 4.24).

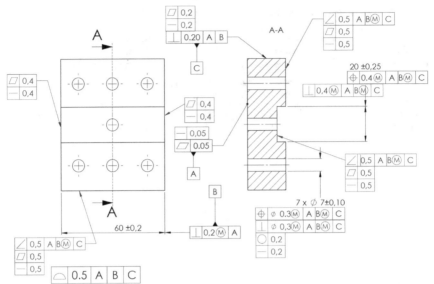

This drawing should be interpreted in accordance with ASME Y14.5-2018
Dimensions not shown on drawing are controlled by means of a 3D CAD file

Fig. 4.24 The blue coloured frames show the implicit indications of the drawing and point out the potentiality of the ASME specification, which allows both the dimensional and geometrical general tolerances to be eliminated, thus making the control phase univocal and coherent. The general tolerance on a profile locates and orientates all the surfaces of the workpiece

4.6   The Main Differences Between ISO GPS and ASME GD&T Standards

Today, the importance of international standards in the technical documentation field is growing at the same rate as the globalisation of production; a simple, clear, univocal and concise three-dimensional description of the designed components is therefore necessary.

In light of this, the adoption of the ISO GPS or ASME standard can surely be the right path to follow in response to the specific requirements of each single company, in order to offer an unambiguous normative context, with simple and coherent rules, thus eliminating uncertainties and confusion in the design, manufacturing and verification phases.

It therefore appears opportune to underline the main differences between the ISO GPS and ASME standards [3].

1. The main difference between the two standards concerns the concept of interdependence between form and dimension, for which ISO uses the **Independency Principle** concept by default, while ASME uses the **Envelope Requirement** (Rule #1, Fig. 4.25).

Fig. 4.25 The effect of two different default principles as outlined in the ISO and ASME standards. The dimensional and geometrical tolerances in the ISO drawing lead to an extreme boundary condition of 10,4 mm (10 + 0,4) and the roundness error may be greater than the dimensional tolerance. In the ASME drawing, a roundness error is already controlled by the dimensional tolerance, and the use of the roundness control therefore only has the purpose of limiting the error

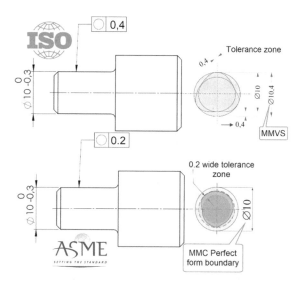

2. A Gaussian dimensioning concept is adopted by default in the ISO standards as a way of specifying the dimension of a feature; in other words, all the points on the surface should fall within an envelope that is obtained by means of the least squares method (this means that some of the points might lie outside a boundary that is defined by the size specification, although it still conforms to the size specification). The **mating size** concept is instead adopted in the ASME standards to describe the size of a feature, that is, all the points on a surface should fall within an envelope which has the maximum material dimensions (Fig. 4.26).

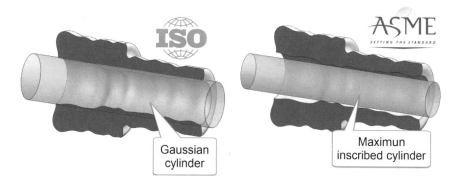

Fig. 4.26 The mating size concept is used in the ASME standards to specify a dimension of a feature (the diameter of the largest inscribed cylinder), while the Gaussian dimensioning concept is utilised in the ISO GPS standards to specify a dimension of a feature

3. The ISO standards express a "**duality**" between specification and verification. Whatever is done in the specification process is mirrored in the actual measurement process. This is described as the "**duality principle**". The ASME standard intentionally distanced itself from the measurement and control process. In fact, in clause 1.6 of ASME Y14.5–2018, the standard states that "*This document is not intended as a gaging standard. Any reference to gaging is included for explanatory purposes only. For gaging principles, see ASME Y14.43 Dimensioning and Tolerancing Principles for Gages and Fixtures*". In other words, the ASME standard describes the acceptable geometry of a part, and not how the part might be measured.

4. The ISO standard is described as "**CMM friendly**", that is, the preferred control system is the coordinate measurement machine. The ideal or nominal geometry is defined in the design process, and the resulting workpiece is the real one in the manufacturing process. The control phase "extracts" the geometry of the physical workpiece in order to elaborate the geometrical features associated with the surfaces (planes, spheres, cylinders) and estimates the value of the measurement of interest (Fig. 2.2). The ASME standard is based on the idea of specifying the geometrically perfect zones within which the real surfaces should be found. This is often referred to as a preference for "**hard gauging**", that is, it is possible to construct functional gauges that represent a physical representation of the tolerance zone (Fig. 4.27).

5. About 150 distinct standards have been issued in the ISO ambit to define the GPS language, while, since back in the 60′s, only a single standard has been developed to define the fundamental rules for the functional dimensioning of components, with the objective of creating a well-defined and coherent normative system, which gave rise to the ASME Y14.5 standard of 2018 (328 pages, Fig. 4.28).

6. Another important difference pertains to the **stability** of the standards: the ISO standard is in continuous evolution, with many changes, sometimes contradicting the previous standards: starting from 2010, the number of standards has grown

Fig. 4.27 Functional gauge used to verify the location of the 4 holes of a plate; the gauging reproduces the worst mating conditions

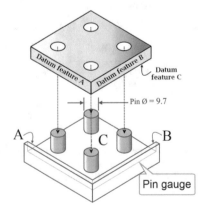

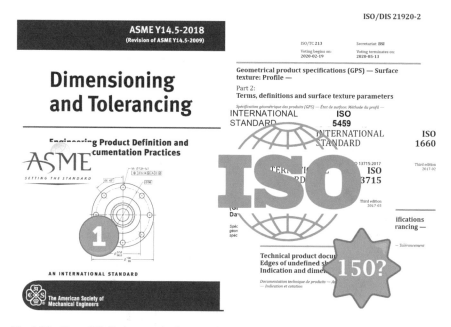

Fig. 4.28   About 150 distinct standards are available in the ISO ambit to define the GPS language, while only a single standard has been developed in the ASME ambit to define the fundamental rules for the functional dimensioning of the components

enormously. Instead, in the ASME ambit, a new GD&T standard is issued every 10–15 years.

7. The tolerances of position and orientation of a feature of size in the ISO standards *are applied to an extracted median line and surface*, while, in ASME, they are applied to ideal (of derived) features, such as axes and centreplanes. For this reason, it is recommended to always indicate the reference standard in the technical drawings of companies, in that, as can be seen in the case of the parallelism tolerance shown in Fig. 4.29, the control of the error leads to two different results: control of the derived median line in the ISO standards and control of the axis in the ASME standard [4].

8. Another non-negligible difference is the use of a comma as the decimal separation in the ISO standards and a dot in the ASME standards (Fig. 4.30).

9. Finally, in the ISO standards, use is made indifferently of the control of a location or profile to locate and orientate a surface, while in the ASME standards, the location tolerance is only used for features of size (Fig. 4.31).

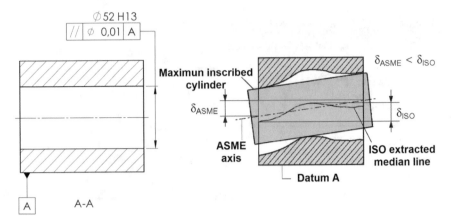

Fig. 4.29 The control of the parallelism tolerance leads to two different results in the ISO and ASME standards

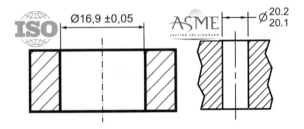

Fig. 4.30 A dot is utilised in the ASME standards as a decimal separation, while a comma is used in the ISO standards

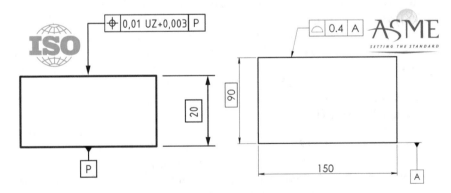

Fig. 4.31 The ASME standards specify that the symbol of location tolerance should only be utilised for features of size (that is, axes and median planes), while the ISO standards allow it to be used to locate surfaces

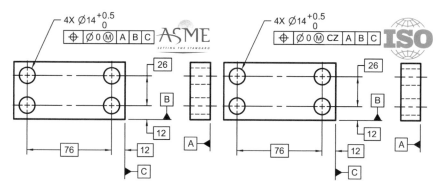

Fig. 4.32 In the ASME standard, the "4x" specification defines a tolerance zone pattern with orientation and position constraints. In ISO, the "4x" specification does not form a pattern. In order to create a tolerance zone pattern specification, it is necessary to use a CZ modifier

4.6.1 The Different Concepts of Pattern and Simultaneous Requirement

The ISO 5458:2018 standard establishes complementary rules to ISO 1101 for pattern specifications. According to the independency principle, a geometrical specification that applies to more than one single feature, by default also applies to those features independently. The tolerance zones defined by one tolerance indicator or by several tolerance indicators should by default be considered independently (this corresponds to the meaning of the modifier SZ, separate zone). In ISO, the 4 holes in Fig. 4.32 with the "4x" specification *do not establish a pattern*. In order to create a tolerance zone pattern specification, it is necessary to use *a tolerance zone pattern modifier*, such as CZ.

The use of the "**Simultaneous requirement**" concept transforms a set of more than one geometrical specification into a combined specification, i.e. a pattern specification.

In the ASME standard, a simultaneous requirement applies to position and profile tolerances that are located by means of basic dimensions related to common datum features *referenced in the same order of precedence at the same boundary conditions*. No translation or rotation takes place between the datum reference frames of the included geometric tolerances for a simultaneous requirement, **thus a single pattern is created**. If such an interrelationship is not required, a notation, such as **SEP REQT**, should be placed adjacent to each applicable feature in the control frame. Figure 4.33 shows an example of the simultaneous requirement principle applied to a two-hole pattern.

According to the ISO 5458:2018 standard, if the geometrical specification has to simultaneously apply to features with location and orientation constraints between the tolerance zones, it is necessary to use tolerance zone pattern modifiers, such as CZ, CZR or SIMn (Fig. 4.34).

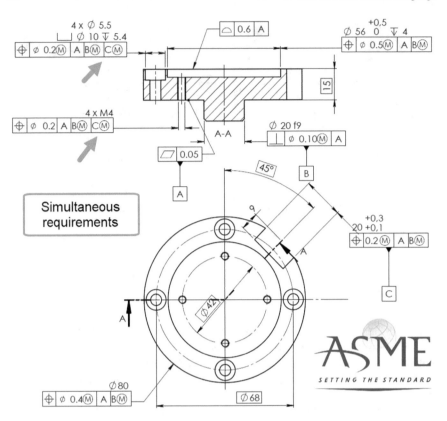

Fig. 4.33 The two pattern tolerance zones in the ASME drawing are simultaneously contained within tolerance zone frameworks related to the same Datum Reference Frame and are thus basically aligned

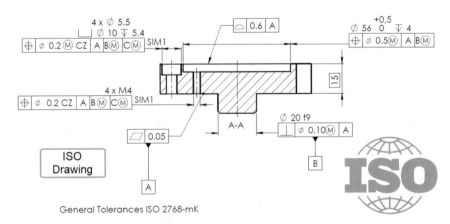

Fig. 4.34 In order to obtain the same functional requirements as the previous figure, the ISO standard uses CZ and SIM1 modifiers

Table 4.3 Comparison of ISO-ASME

	ISO	ASME
Boundary concept	Gaussian least square size	Mating size
Specification versus verification	Duality between specification and verification	Describes the acceptable geometry of a part, not how the part might be measured
Inspection&control	CMM friendly	Hard gauging
Default	Independency principle	Envelope requirement
Axis and centerplane concept	Position and orientation of FOS (RFS) applied to an extracted median line and surface	Position and orientation of FOS (RFS) applied to an ideal feature, such as an axis or centerplane
Stability	Many changes every year, introduction of new concepts and ideas	Revised on average every 10–15 years
Rules versus examples	Based only on rules	Based on rules and examples
Pattern concept	To create a tolerance zone pattern specification, it is necessary to use a tolerance zone pattern modifier 2x Ø0,2 CZ Ⓜ	The "nx" specification defines a pattern 2x Ø0,2 Ⓜ
Simultaneous requirement	The pattern modifiers CZ, CZR and SIMn establish a simultaneous requirement	A common datum feature reference establishes a simultaneous requirement

Table 4.3 summarises the main differences between the two standards.

References

Effenberger G (2013) Geometrical Product Specifications (GPS)—consequences on the tolerancing of features of size, TEQ training & consulting GmbH.

Day D (2009) The GD&T Hierarchy (Y14.5 2009), Tec-Ease, Inc.

Morse E (2016) Tolerancing standards: a comparison. Quality Magazine

Henzold G (2006) Geometrical dimensioning and tolerancing for design, manufacturing and inspection: a handbook for geometrical product specification using ISO and ASME standards. Butterworth-Heinemann

Chapter 5
Interdependence Between Dimensions and Geometry

Abstract In ISO GPS terminology, the maximum material requirement, (MMR) and least material requirement (LMR) represent two of the fundamental rules on which geometrical dimensioning with tolerances is based, and which are the subject of the ISO 2692 standard. The designer, when establishing a maximum or least material requirement, defines a geometrical feature of the same type and of perfect form, which limits the real feature on the outside of or inside the material. MMR is used to control the assemblability of a workpiece while LMR is used to control a minimum distance or a minimum wall thickness. This chapter also introduces the Reciprocity requirement (RPR) and the "zero tolerance" concepts and offers practical examples to guide a designer in his/her choice of the correct requirement from the geometrical tolerance specifications.

5.1 MMR and LMR Requirements

As already seen, the ISO 8015 standard has established the independency principle between dimensional tolerances and geometrical tolerances; however, there are some exceptions contained in the ISO 2692 standard (current version ISO/DIS 2692:2019) which, by using the **maximum material requirement** (MMR) and **least material requirement** (LMR), introduce an interdependency between dimensions geometry. The MMR and LMR requirement allow two independent requirements to be combined into one collective requirement, or a maximum material or least material virtual condition to be defined directly, in order to simulate the intended function of the workpiece.

The geometrical tolerance is in fact considered to have been applied, regardless of the size of the workpiece (in ASME *Regardless of Feature Size*, RFS), each time an exception is not specified (utilising a symbol named modifier), which could be the requirement an envelope Ⓔ, or the application the MMR Ⓜ and/or LMR Ⓛ requirements.

In the ISO standard, a distinction is made between:

1. *Maximum Material Condition*, **MMC**, the **state** of the considered extracted feature, where the feature of size is at that limit of size where the material of

the feature is at its maximum everywhere, e.g. minimum hole diameter and maximum shaft diameter.

2. *Maximum Material Size*, **MMS**: the dimension that defines the maximum material condition of a linear feature of size.

3. *Least Material Condition*, **LMC**, the **state** of the considered extracted feature, where the feature of size is at that limit of size where the material of the feature is at its minimum everywhere, e.g. maximum hole diameter and minimum shaft diameter

4. *Least Material Size*, **LMS**, the dimension that defines the least material condition of a feature of size.

5.1.1 Maximum Material Requirement (MMR)

Sometimes called the *Maximum Material Principle* but, in the ISO GPS terminology, correctly defined as condition or requirement, the *Maximum material requirement*, (MMR) represents one of the fundamental rules on which geometrical dimensioning with tolerances is based, and which is the subject of the ISO 2692 standard.

The mating characteristics of the workpieces that must be assembled depend on the joint effect of the actual dimensions, and of the form and location errors of the features that have to be mated.

In the case of a mating with clearance, the minimum assembly clearance is obtained when each of the mating features of size is at its maximum material size (e.g. the largest bolt size and the smallest hole size) and when the geometrical deviations (e.g. the form, orientation and location deviations) of the features of size and their derived features (median line or median surface) are also fully consuming their tolerances.

The assembly clearance increases to a maximum when the sizes of the assembled features of size are furthest from their maximum material sizes (e.g. the smallest shaft size and the largest hole size) and when the geometrical deviations (e.g. the form, orientation and location deviations) of the features of size and their derived features are zero.

The consequence is that the prescribed tolerances *can in practice be enlarged*, without compromising the possibility of mating, when the actual dimensions of the features that have to be mated do not reach the values corresponding to the maximum material condition (Fig. 5.1).

This constitutes the MMR principle, or maximum material requirement, and it is indicated on a drawing with the symbol Ⓜ. This symbol is inserted after the value of a tolerance in a tolerance frame indicator, and should be read as "*the geometrical tolerance here imposed and foreseen for the case in which the linear dimensions are under the maximum material condition*". The foreseen tolerances can therefore be increased by a value[1] that is equal to the difference between the dimension of the maximum material and the actual dimension (Fig. 5.2). An increase in the geometrical

---

[1]This increase is generally called **bonus** in the ASME standard.

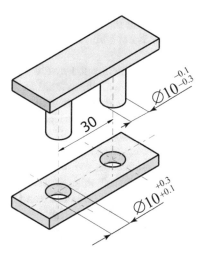

Fig. 5.1 The mating of a plate with two through holes and a feature with two pins; the most critical conditions pertaining to the distance between their centres occur for the maximum material condition, that is, when the hole is at the minimum diameter (Ø10.1) and the pin is at the maximum diameter (Ø9.9). It is obvious that when the hole is at a diameter of 10.3 mm and the pin is at a minimum diameter of 9.7 mm, the tolerance on the distance between their centres could be increased without jeopardising the mating

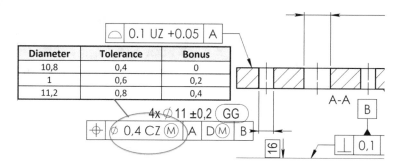

Fig. 5.2 An increase in the location tolerance (Bonus) due to the presence of the Ⓜ modifier for the component shown in Fig. 3.1. The increase is equal to the difference between the measured dimension and the maximum material dimension

tolerance can obviously be applied when the feature to which the MMR requirement has to be applied is a feature of size (to which a dimensional tolerance can be associated), with an axis or a symmetry plane, such as a hole, a groove or a pin. The advantages that can be obtained pertain to an economy in the production, as a result of the enlargement of the limits of the tolerances, and a reduction in the wastes, since it is possible to accept features which, although the geometrical tolerances are not within the prescribed limits, in practice offer the same functional characteristics as the features achieved within the limits. An increase in the location tolerance can in general be accepted, for example, for the distances between the centres of holes for bolts, coach screws, etc., while it is not admissible for the axes of gears, dowel pins, kinematic connections, etc.

In short, the assembly function is controlled by the maximum material requirement, which is indicated on drawings with the symbol Ⓜ.

The designer, when establishing a maximum material requirement, defines a geometrical feature of the same type and of perfect form of a value that is equal to the **Maximum Material Virtual Size** (MMVS), which limits the real feature on the *outside* of the material. In this case, the MMR i*s used to control the assemblability* of the workpiece, and the real feature cannot violate the MMVS.

The *Maximum Material Virtual Size* is the size generated by the collective effect of the maximum material size (MMS) of a feature of size and the geometrical tolerance (form, orientation or location) given for the derived feature of the same feature of size. The **Maximum Material Virtual Condition** (MMVC) is the **state** of an associated feature of the maximum material virtual size, MMVS.

A shaft subjected to dimensional tolerances on the diameter is shown in Fig. 5.3, for which a straightness tolerance of 0.1 mm is indicated. In this case, the tolerance refers to the straightness of the derived median line, since the symbol is placed on the diameter dimension. If the maximum material requirement is not indicated, the value of the straightness tolerance remains constant, since the value of the diameter varies from a maximum to a minimum, as a result of the independency principle. In this case, the straightness tolerance always imposes that the derived median line should always fall within a cylindrical tolerance zone of 0.1 mm.

When an MMR symbol is included, the intended function of the shaft indicated in Fig. 5.3 could be a clearance fit with a hole of the same length as the toleranced cylindrical feature. In this case, the extracted feature should not violate the maximum

Fig. 5.3 Application of MMR for an external cylindrical feature on the basis of size and form (straightness) requirements

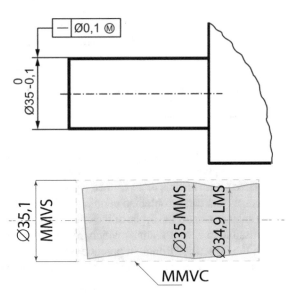

material virtual condition, MMVC, which has the MMVS diameter = 35,1 mm, since MMVS is obtained with the following formula for an external feature of size:

$$MMVS = MMS + \delta, \text{ where } \delta \text{ is the geometrical tolerance.}$$

At the same time, the extracted feature should have a larger local diameter than LMS = 34.9 mm and a smaller one than MMS = 35.0 mm. The orientation and location of the MMVC are not controlled by any external constraints.

The designer, through the implicit indications of the MMVC, is able to ensure full functionality and interchangeability of the produced workpieces, all at a minimum cost. The 35 mm hole in the plate in Fig. 5.4 has an MMVC of 35.1 mm, which is obtained by subtracting the tolerance of 0.1 mm from the maximum material size (MMS) of 35.2 mm. For internal features of size, MMVS is obtained from the formula:

$$MMVS = MMS - \delta, \text{ where } \delta \text{ is the geometrical tolerance.}$$

The intended function of the hole could be for assembly with a pin shaft, where the functional requirement is that the two planar faces should be in contact and, at the same time, the pin should fit into the hole. The extracted feature should not violate the maximum material virtual condition, MMVC, which has the MMVS diameter = 35.1 mm and the extracted feature should simultaneously have a smaller local diameter than LMS = 35.3 mm and larger than MMS = 35.2 mm. MMVC is orientated perpendicular to the datum and the location of the MMVC is not controlled by any external constraints. In short, the boundaries of the hole must not violate the MMVC cylindrical boundary.

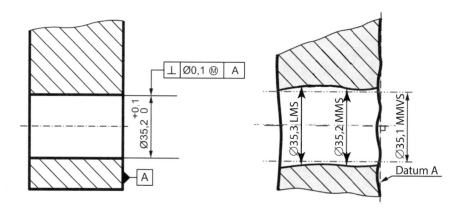

Fig. 5.4 An application of MMR for an internal cylindrical feature on the basis of size and orientation requirements

5.1.2   Least Material Requirement (LMR)

A least material requirement is designed to control, for example, the minimum wall thickness, thereby preventing breakout (due to pressure in a tube), the maximum width of a series of slots, etc. It is indicated on drawings with the symbol Ⓛ In the case in which the *Least Material Requirement*, LMR, is foreseen, it is not difficult to understand why reference is made to holes which have the maximum admissible diameter according to the prescribed dimensional tolerance, or to minimum diameter shafts.

Again, in this case, when different conditions from the LMC are available, an increase in the form and location tolerances can be made that is equal to the difference between the actual dimensions and that of the least material.

In short, the symbol Ⓛ is used on the drawing with the objective of guaranteeing the existence of a resistant section, or of guaranteeing a minimum distance. Figure 5.5 shows a typical application of the LMC requirement introduced to protect the minimum distance of the edge of a hole with respect to the edge of the workpiece, but at the same time to permit an increase in the location tolerance.

Fig. 5.5 Utilisation of the LMR requirement to control a minimum distance

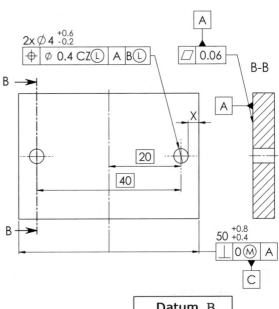

	Datum B	
Hole	LMC	MMC
diameter	50,4	50,8
LMC 4,6	0,4	0,8
4,4	0,6	1,0
MMC 4,2	0,8	1,2

Fig. 5.6 The minimum distance X is protected for any dimensional condition (2.7 mm)

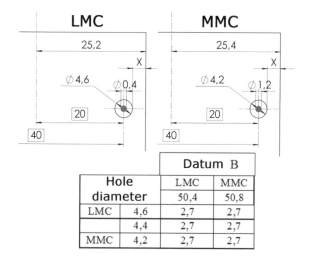

Hole		Datum B	
		LMC	MMC
diameter		50,4	50,8
LMC	4,6	2,7	2,7
	4,4	2,7	2,7
MMC	4,2	2,7	2,7

The worst mating condition occurs when the hole has the maximum allowed dimensions (or when the datum has a minimum width, LMC), and it is in fact in this condition that the localisation tolerance of 0.4 mm is defined. When a hole is produced with smaller dimensions (or when the datum is produced with greater dimensions), it is possible to increase the localisation tolerance up to a value of 1.2 mm; as can be seen in the dimension chain (Fig. 5.6), the distance X remains the same under any condition, in that at the MMC state:

$$X = \frac{50.8}{2} - \left(\frac{40 + 4.2 + 1.2}{2}\right) = 2.7\,\text{mm}$$

While, at the LMC state:

$$X = \frac{50.4}{2} - \left(\frac{40 + 4.6 + 10.4}{2}\right) = 2.7\,\text{mm}$$

The least material requirement can also be characterised by a collective requirement pertaining to the size and the geometrical deviation of the feature of size. When a designer introduces the LMR symbol, he/she defines a geometrical feature of the same type and of perfect form, with a value equal to the **Least Material Virtual Size** (LMVS) which limits the real feature on the *inside* of the material.

The Least Material Virtual Size is the size generated by the collective effect of the Least material size (LMS) of a feature of size and the geometrical tolerance (form, orientation or location) given for the derived feature of the same feature of size. The **Least Material Virtual Condition** (LMVC) is the *state* of an associated feature of maximum material virtual size, LMVS.

The 4 mm hole in the plate in Fig. 5.5 has an LMVC of 5 mm, which is obtained by adding the tolerance of 0.4 mm from the minimum material size (LMS) of 4.6 mm.

For internal features of size, LMVS is obtained from the formula:

$$\text{LMVS} = \text{LMS} + \delta, \text{ where } \delta \text{ is the geometrical tolerance.}$$

The LMR in Fig. 5.7 requires the extracted surface of the hole to fall inside the LMVC boundary.

Figure 5.8 shows an example of coaxiality control with a datum established from an external feature of size referenced at LMC. The least material virtual size (LMVS) of an external feature of size is obtained from the formula:

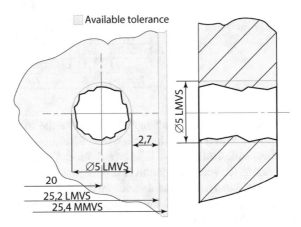

Fig. 5.7 The LMR in Fig. 5.5 requires that the extracted surface of the hole should fall inside the LMVC boundary

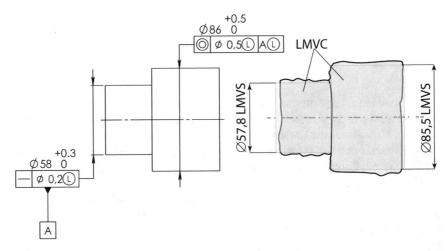

Fig. 5.8 An example of an LMR for an external cylindrical feature based on size and location (coaxiality) requirements with the axis of a cylindrical feature as the datum referenced at LMR

$$\text{LMVS} = \text{LMS} - \delta, \text{ where } \delta \text{ is the geometrical tolerance}$$

LMR requires that the extracted surface of a feature of size should fall outside the LMVC boundary. The location of the LMVC is at 0 mm from the axis of the LMVC of the datum feature. The extracted feature of the toleranced feature should not violate the least material virtual condition, LMVC, which has the diameter LMVS = 85.6 mm. The same extracted feature should have a larger local diameter than LMS = 86 mm and smaller than MMS = 86.5 mm.

The extracted feature of the datum feature should not violate the least material virtual condition, LMVC, which has an LMVS diameter = 57.8 mm. The same extracted feature should have a larger local diameter than LMS = 58 mm and smaller than MMS = 58.3 mm.

5.1.3 Reciprocity Requirement (RPR)

The **reciprocity requirement** (RPR) is an additional requirement, that may be used together with the maximum material requirement and the least material requirement, and is indicated on a drawing with the symbol ®. RPR allows a size tolerance to be enlarged whenever the geometrical deviation on the actual workpiece does not take full advantage of, the maximum material virtual condition or the least material virtual condition, respectively.

RPR is basically an additional requirement for a feature of size and it is introduced, in addition to the maximum material requirement, MMR, or the least material requirement, LMR, to indicate that the size tolerance is increased by the difference between the geometrical tolerance and the actual geometrical deviation.

The reciprocity symbol defines equivalent requirements to specifications with dimensional and geometrical requirements with an indication of zero tolerance with a modifier Ⓜ. A reciprocity requirement should only be indicated in the tolerance section of the tolerance indicator. As stated in the standard, the modifier *shall not be indicated in the datum section* of the tolerance indicator.

RPR allows the distribution of the variation allowance between dimensional and geometrical tolerances to be chosen on the basis of the manufacturing capabilities. It should be noted that this requirement is not present in the ASME standard.

Let us consider the workpiece shown in Fig. 5.9, which has four 9 mm diameter holes and a position tolerance of 0.4 mm; the maximum material virtual size (MMVS) of the hole, as has already been seen, is calculated from the difference 8.8 − 0.4 mm = 8.4 mm, which represents the internal MMVC boundary that cannot be violated.

Therefore, a hole verified with a diameter of 8,4 mm could be mated without any problems if it has a zero-position error, *but the component would nevertheless be discarded* during the dimensional control.

In order to avoid this problem, it is possible to utilise the *zero tolerance method* (Fig. 5.10), that is, the hole is produced in such a way that a position tolerance of zero

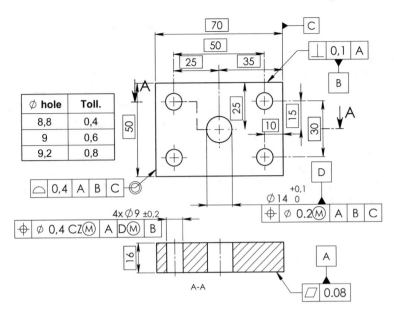

Fig. 5.9 An indication of a position tolerance: the table shows the variations of the tolerance as the diameter of the hole varies as a result of utilising the maximum material requirement, MMR

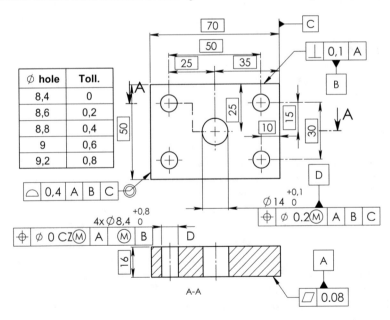

Fig. 5.10 An indication of zero tolerance. As can be seen in the table, the tolerance is only zero if the hole is produced at the maximum material condition (8.4 mm) and increases up to 0.8 mm

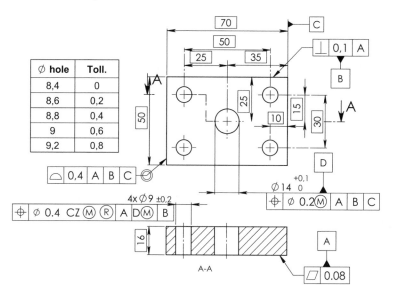

Fig. 5.11 The drawing above has the same dimensions as Fig. 5.9, but the reciprocity symbol makes it possible for the allowed variations between the dimensional and geometrical errors to be distributed. In practice, the same effect as a zero tolerance is obtained

is obtained for the virtual dimension but, at the same time, the dimensional tolerance zone is increased.

Several advantages can thus be obtained, including:

(a) full use of the dimensional tolerance bonus (from 0 to 0.8 mm);
(b) the possibility of controlling the position and the lower dimensional limits of the hole at the same time with a functional gauge;
(c) a reduction in the production costs, since reusable parts are not rejected.

The additional requirement of RPR in Fig. 5.11 changes the size tolerance of the hole as a result of adding the collective MMR requirement, thereby obtaining the same effect as the indication of a zero tolerance, without any change in the drawing.

From the functional point of view, that is, as far as assembly problems are concerned, the three specifications are equivalent; from the production point of view, undoubtable advantages can be obtained.

In this sense, the allowable positional tolerance values are listed on the vertical scale in the analysis chart in Fig. 5.12. The horizontal scale shows the virtual condition and hole sizes of the part. In the case in which the zero tolerance is not used, the cyan coloured area represents the zone of the workpieces accepted during the control according to the drawing specifications, while the parts that fall into the yellow area are rejected during the control, even though they are acceptable from the functional point of view.

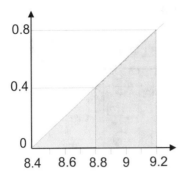 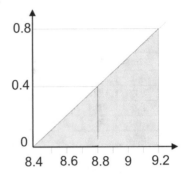

Fig. 5.12 Tolerance analysis charts for the component dealt with in Fig. 5.9; the workpieces in the chart on the left that fall into the yellow zone are rejected during the control, even though they are acceptable from a functional point of view; in the case of the use of a zero tolerance or RPR (chart on the right), the cyan area in the chart becomes larger, thus allowing a reduction in waste and more manufacturing flexibility to be achieved

When the zero-tolerance concept, or the reciprocity requirement, is adopted, the cyan area in the chart becomes larger, thus allowing a reduction of the rejected parts and greater manufacturing flexibility.

5.2   Direct Indication of Virtual Size

The new ISO 2962 standard allows a designer to directly specify the value of the maximum virtual size or the least material virtual size. In this case, the calculated value of the virtual size should be indicated between brackets in the tolerance indicator. If a size is also specified for the considered feature, it should be considered as an independent specification according to ISO 14405–1. No collective requirement is created between the two specifications (size specification and geometrical specification) in the case of the direct indication of the maximum material or least material virtual size.

When the maximum material requirement, MMR, applies to the tolerance feature, and the direct indication of virtual size is selected, then the value of the maximum material virtual size (MMVS) should be indicated after the Ⓜsymbol between brackets (Fig. 5.13). The Ø symbol should be omitted when the feature of size is not a diameter. When a direct indication of the maximum material virtual size (MMVS) is selected, no geometrical tolerance value should be indicated before the symbol. The same rule applies to the LMR requirement (Fig. 5.14).

When the maximum material requirement *applies to a datum* and the direct indication of virtual size is selected, then the maximum material virtual size (MMVS) for the datum should be indicated with a numerical value between brackets in the datum indicator, after the datum feature symbol, as shown in Fig. 5.15. The Ø symbol should be omitted when the feature size is not a diameter. It is necessary to point out

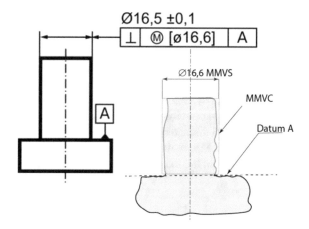

Fig. 5.13   Direct indication of an MMVS on an external feature

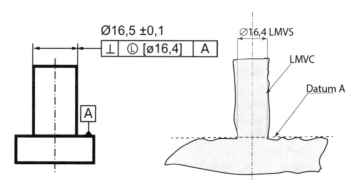

Fig. 5.14   Direct indication of an LMVS on an external feature

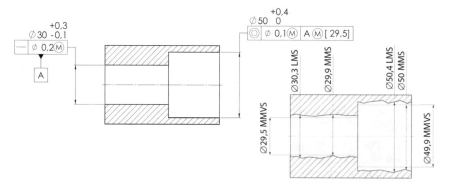

Fig. 5.15   Example of a direct indication of an MMVS on a datum

that, for the interpretation of straightness, the extracted feature should not violate the maximum material virtual condition, MMVC, which has an MMVS diameter = 29.9 − 0.2 = 29.7 mm. The extracted feature should have a smaller local diameter than LMS = 29.9 mm and larger than MMS = 30.3 mm. The extracted feature of the datum feature should not violate the maximum material virtual condition, MMVC, which has an MMVS diameter = 29.5 mm. At the same time, the extracted feature of the toleranced feature should not violate the maximum material virtual condition, MMVC, which has an MMVS diameter = 49.9 mm. The same feature should have a smaller local diameter than LMS = 50.4 mm and larger than MMS = 50 mm. The MMVC is located at 0 mm from the axis of the MMVC of the datum feature.

5.3   When to Use MMR, LMR, and RFS

The decision to choose an RFS, MMR, LMR or RPR requirement depends on the function of the workpiece. MMR is generally chosen when the parts have to be assembled with a mating with clearance (i.e. fixed fastener), but it is useful to allow additional maximum tolerances to reduce the manufacturing costs.

LMR is useful when a designer has the goal of guaranteeing a minimum distance or a critical wall thickness. RPR is used to enlarge the size tolerance, for example, in a non-functional assembly. Finally, RFS (without any modifier) is used when any additional tolerance, such as the control of a position tolerance of a locating pin, can compromise the functionality of the workpiece.

It should be noted that, in many cases, an additional tolerance does not have any impact on the function, and therefore does not make the part work better. The additional variation allowed by MMR may lead to a reduction in the cost of manufacturing parts, as it may allow certain functional parts to pass inspection.

Figure 5.16 in fact shows a typical application in which it is appropriate to use the MMR to guarantee the correct position of a sensor, which is positioned by means of a screw. These parts have a fixed fastener mating, that is, one of the parts has a threaded hole, and the other has a clearance hole. In this case, the position of the sensor holder only depends on the threaded hole. The screw has a mating clearance with the plate (Fig. 5.17), and the clearance hole position error therefore does not affect the position of the sensor. In this case, any additional tolerance does not have an impact on the function, and it is convenient to specify the *maximum material requirement*.

In the second case in Fig. 5.18, the sensor support is only positioned by means of a dowel pin, which allows mating with only a slight interference with the plate. However, the error in locating the hole in the plate will affect the position of the sensor, and it is therefore better not to further increase the amount of tolerance through the use of an MMR modifier (Fig. 5.19).

In the latter case, the connection of the sensor support is obtained with a pin from the same support, which is connected, with a clearance fit, to the plate (Fig. 5.20). The worst condition, which affects the location of the sensor, occurs when the hole has the

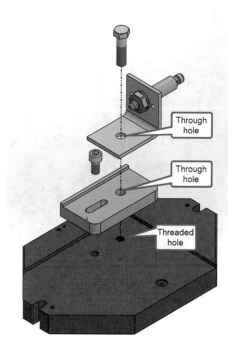

Fig. 5.16 A fixed fastener assembly: the designer has the goal of guaranteeing the correct position of a sensor

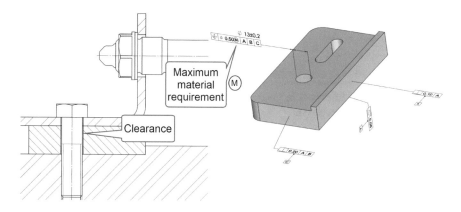

Fig. 5.17 The clearance hole position error does not affect the position of the sensor

maximum allowed dimensions LMS (and when the pin has the minimum allowed diameter LMS). In such a case, it is useful to use the LMR modifier. Figure 5.21 shows that when assuming an MMVS mating pin size of 12.8 mm and accepting a sensor position error of less than 1 mm, the hole is produced with a smaller size,

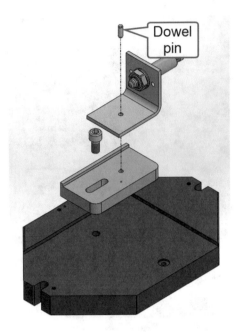

Fig. 5.18 The sensor support is only positioned by means of a dowel pin

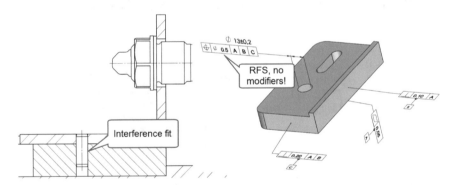

Fig. 5.19 A dowel pin which provides a mating with only a slight interference with the plate. The error in locating the hole in the plate will affect the position of the sensor, and it is therefore better not to further increase the amount of tolerance through the use of an MMR modifier

and it is possible to increase the localisation tolerance so that the LMR requirement guarantees the *minimum position error* and the *maximum admissible tolerance*.

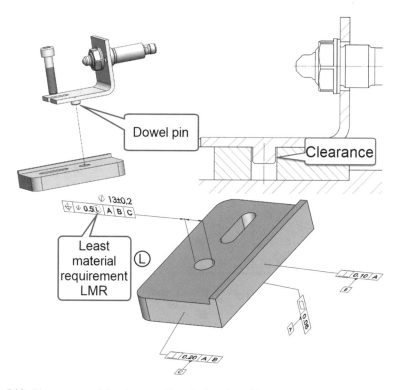

Fig. 5.20 The worst condition that can affect the location of the sensor, occurs when the hole has the maximum allowed dimensions LMS (and when the pin has the minimum allowed diameter, LMS); it is therefore useful to use an LMR modifier

5.4 The MMR and LMR Requirements in the ASME Standards

The ASME standard does not use the term "requirement" to specify an interdependence between size and geometric tolerance, but, depending on its function, the form deviation of a feature of size is always controlled by its size and any applicable geometric tolerances.

In the ASME Y14.5:2018 standard, the modifiers M and L are introduced in Sect. 5.9 through Rule #2: "*RFS is the default condition for geometric tolerance values. The Maximum Material Condition (MMC) or Least Material Condition (LMC) modifier may be applied to a geometric tolerance value to override the RFS default*".

The concept of Maximum Material Condition (MMC) or Least Material Condition (LMC) is used to describe the size limit of a feature of size at which the part contains the maximum or minimum amount of material. An MMC size limit combined with Rule #1 describes a boundary of perfect form at the MMC.

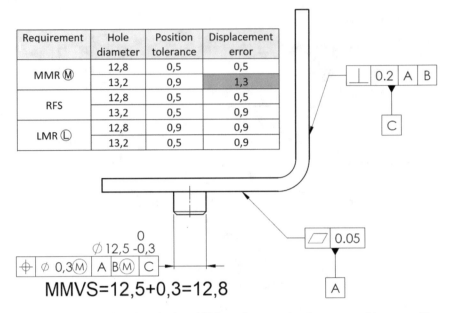

Requirement	Hole diameter	Position tolerance	Displacement error
MMR Ⓜ	12,8	0,5	0,5
	13,2	0,9	1,3
RFS	12,8	0,5	0,5
	13,2	0,5	0,9
LMR Ⓛ	12,8	0,9	0,9
	13,2	0,5	0,9

MMVS=12,5+0,3=12,8

Fig. 5.21 If an MMVS mating pin size of 12.8 mm is assumed and a sensor position error of less than 1 mm is accepted, then the LMR requirement guarantees the minimum position error and the maximum admissible tolerance

In the GD&T language, a set of symbols, called "**modifiers**", is used to communicate additional information about the drawing or tolerancing of a part. The MMC Ⓜor LMC Ⓛ*material condition modifier* may be applied to a geometric tolerance value *to override the RFS default*.

However, the ASME standard distinguishes the MMC (or LMC) specification applied to a feature of size from the MMB (Maximum Material Boundary) (or Least Material Boundary, LMB) specification applied to each datum feature reference.

MMB is the worst-case boundary that exists on or *outside* the material of a feature(s) and it is obtained as a result of the combined effects of size and geometric tolerances.

LMB is the worst-case boundary that exists on or *inside* the material of a feature(s) and it is obtained as a result of the combined effects of size and geometric tolerances.

RFS is the default condition for geometric tolerance values, while RMB (*Regardless Material Boundary*) is the default condition for a datum. An MMB or LMB material boundary modifier may be applied to a datum feature reference to override an RMB default.

Consistently with the ISO specification (see Table 5.1), circular run-out, total run-out and orientation tolerances applied to a surface, profile of a line, profile of a surface, circularity or cylindricity cannot be modified to apply an MMC or LMC.

When a geometric tolerance, applied to a feature of size, is specified with an MMC or LMC modifier, a constant boundary (named **Virtual Condition**, VC) is generated

Table 5.1 Applicability of MMR or LMR with the various characteristic geometrical symbols

Specification	Characteristics	Symbol	Applicability Ⓜ Ⓛ
Form	Straightness	—	Yes
	Flatness	▱	Yes
	Roundness	○	No
	Cylindricity	⌭	No
	Line profile	⌒	No
	Surface profile	⌓	No
Orientation	Parallelism	//	Yes
	Perpendicularity	⊥	Yes
	Angularity	∠	Yes
	Line profile	⌒	No
	Surface profile	⌓	No
Location	Position	⊕	Yes
	Coaxiality (Concentricity has been removed from the ASME Y14.5:2018 standard.)	◎	Yes
	Symmetry (Symmetry has been removed from the ASME Y14.5:2018 standard)	⌟	Yes
	Line profile	⌒	No
	Surface profile	⌓	No
Run-out	Circular run-out	↗	No
	Total run-out	↗↗	No

from the collective effects of the MMC or LMC modifier and the geometric tolerance of that material condition. This virtual condition (or **Worst-Case Boundary**,
WCB) is the extreme boundary that represents the worst-case for such functional
requirements, such as clearance, assembly with a mating part, thin wall conservation
or hole alignment.

Figure 5.22 shows the collective effects of MMC and the applicable tolerances.
The increase in size of the position tolerance zone in the table is commonly referred
to as "**bonus tolerance**" in ASME and is viewed by many to be "*extra tolerance for
free*".

Figure 5.23 illustrates the concept of **inner boundary**, a worst-case boundary
generated from the collective effects of the smallest feature of size (MMC for an
internal feature of size, LMC for an external feature of size) and the applicable
geometric tolerance. An **outer boundary** is a worst-case boundary generated from
the collective effects of the largest feature of size (LMC for an internal feature of
size, MMC for an external feature of size) and the applicable geometric tolerance.

The **resultant condition** (RC) is the *single* worst-case boundary generated from
the collective effects of a specified MMC or LMC of a feature of size, the geometric
tolerance of that material condition, the size tolerance, and the additional geometric
tolerance derived from the departure of the feature from its specified material
condition.

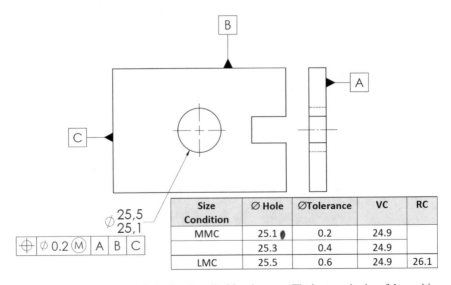

Size Condition	Ø Hole	ØTolerance	VC	RC
MMC	25.1 ●	0.2	24.9	
	25.3	0.4	24.9	
LMC	25.5	0.6	24.9	26.1

Fig. 5.22 Collective effects of MMC and applicable tolerances. The increase in size of the position
tolerance zone in the table is commonly referred to as "bonus tolerance" in ASME

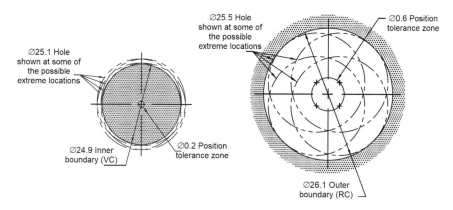

Fig. 5.23 Virtual and Resultant Condition Boundaries for an internal feature, using an MMC modifier

Figures 5.24 and 5.25 show the same concepts applied to an external feature, that is, a plate with two pins. By using the inner and outer boundary values, it is possible to calculate the minimum and maximum distances between the two pins measured with a gauge (Fig. 5.26).

As seen in the ISO standard, in cases where the MMB or LMB boundary is not clear, or another boundary is required, the value of a fixed-size boundary can be indicated, enclosed in brackets, following the applicable datum feature reference and any modifier (Fig. 5.27).

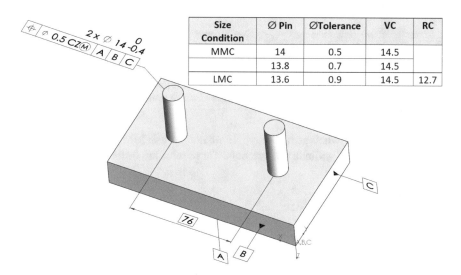

Size Condition	Ø Pin	ØTolerance	VC	RC
MMC	14	0.5	14.5	
	13.8	0.7	14.5	
LMC	13.6	0.9	14.5	12.7

Fig. 5.24 Collective effects of MMC and the applicable tolerances for an external feature

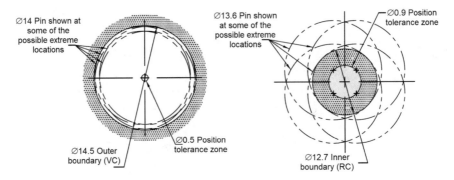

Fig. 5.25 Virtual and Resultant Condition Boundaries for an external feature, using an MMC modifier

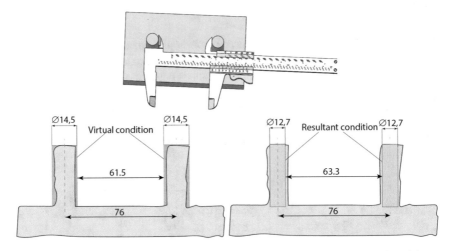

Fig. 5.26 By using the inner and outer boundary values, it is possible to calculate the minimum and maximum distances between the two pins measured with a gauge

Table 5.1 shows whether it is possible or not, to use MMR or LMR with the various characteristic geometrical symbols. These rules are in force in both the ISO and ASME standards.

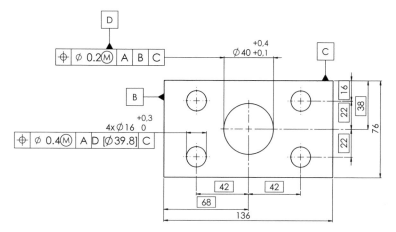

Fig. 5.27 Explicit Specification of True Geometric Counterpart Boundaries in the ASME standard

Chapter 6
Datums and Datum Systems

Abstract A datum system has the purpose of defining a set of two or more ideal features established in a specific order (for example, a system made up of a triad of mutually orthogonal planes) that allows not only the tolerance zones to be orientated and located, but also their origin to be defined for the measurement, and the work-piece to be blocked during the control. When it is not desirable to use a complete integral feature to establish a datum feature, it is possible to indicate portions of the single feature (areas, lines or points) and their dimensions and locations using datum targets. This chapter illustrates the main differences between the ISO 5459 and ASME standards for the specification of datums and highlight some theoretical and mathematical concepts. This section also provides simple rules to follow when-ever choosing functional datums to ensure the part will function as intended, with the least possible amount of variation.

6.1 Datum Systems

A datum system represents one of the most advantageous instruments to pass on information about functional relationships and, at the same time, to transmit the modalities and sequences with which the control of the component should be carried out in such a way that the inspection measurement is univocal and repeatable.

The greatest disadvantage of the use of a datum system is that an adequate knowl-edge of the geometrical tolerance standards is necessary. In fact, a lack of attention to the functional aspects of geometrical dimensioning has led to some datum miscon-ceptions. The first of these concerns the conviction that datums **exist on a part**, although the planes and axes taken as datums are in fact abstract concepts that are obtained through complex verification operations.

Another misconception is that datums are established directly **during the manu-facturing design phase**; in fact, the datums derived from operational requirements often do not agree with the ones that are fixed during the design phase on the basis of the functional requirements. If a process engineer introduces any modifications, the functional requirements will no longer be respected, because of changes in the tolerances (which often become more reduced).

S. Tornincasa, *Technical Drawing for Product Design*, Springer Tracts
in Mechanical Engineering, https://doi.org/10.1007/978-3-030-60854-5_6

A datum should always be selected on the basis of the functions of a part, since choosing datums on the basis of the technological location of the workpieces can lead to a reduction in the available tolerances.

A datum system (known as DRF, *Datum Reference Frame* in the ASME standards) is a set of two or more ideal features established in a specific order from two or more datum features, with the aim of (Fig. 6.1):

1. defining *the origin* from which the location or geometric characteristics of the features of a part are established;
2. allowing the workpiece *to be blocked* during a control, so that the control is univocal and repeatable;
3. *locating* and *orienting* the tolerance zones.

In short, datums allow a datum system to be defined for the dimensioning of a drawing and they can be considered as a way of **blocking the degrees of freedom of a tolerance zone**.

The ISO 5459 standard specifies the terminology, rules and methodologies pertaining to the indication and comprehension of datums and of datum systems

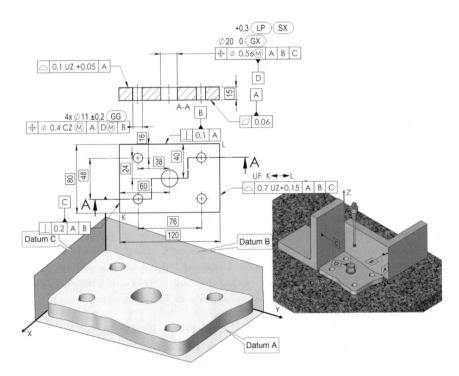

Fig. 6.1 A datum system has the purpose of defining a set of two or more ideal features, established in a specific order (for example, a system made up of a triad of mutually orthogonal planes), that allows not only the tolerance zones to be orientated and located, but also their origin to be defined for the measurement, and the workpiece to be blocked during the control

in the technical documentation of a product. The physical surfaces of real and imperfect parts are defined as **datum features** in the standard as they serve the purpose of constraining the rotational and translational degrees of freedom during the assembly processes and which can be used to define the datums (Fig. 6.2). In fact, each workpiece is considered to be made up of features that can have a size and an axis, or a symmetry plane (a feature of size such as a groove, a hole or a pin), or non-dimensional features such as flat or cylindrical surfaces; as each feature can be considered as a datum, it is possible to have datums correlated to both sizeable and non-sizeable features, which are indicated on a drawing in different ways.

A datum is a theoretically exact point, straight line, or plane from which the location and orientation of features, or both, can be defined (Fig. 6.3). A datum specification refers to the concept of a **situation feature**, which is obtained through

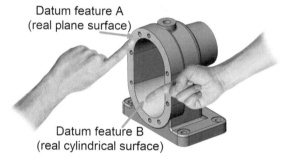

Fig. 6.2   Datum features are particular physical features of real, imperfect, labelled workpieces that have the purpose of constraining the rotational and translational degrees of freedom of relative parts. Datums are the surfaces of the workpiece that can be touched, and they are suitable for associating to a datum

Fig. 6.3   A datum is a theoretically exact point, straight line, or plane from which the location and orientation of features, or both, can be defined

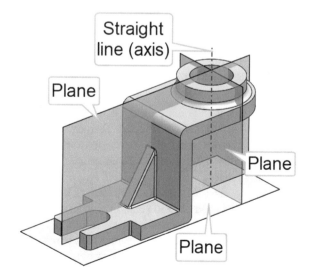

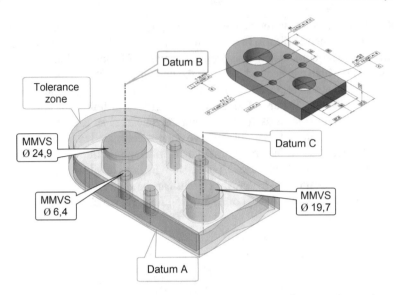

Fig. 6.4 According to ISO 5459, a datum is one or more situation features (plane, point and/or straight line) that is selected to define the location or orientation, or both, of a tolerance zone or an ideal feature that represents, for instance, a virtual condition

an association criterion with the datum feature, that is, an ideal feature (plane, point or straight line) is associated with the true (extracted and filtered) surface.

In short, a datum is one or more *situation features* of one or more features *associated* with one or more real integral features selected *to define the location or orientation, or both, of a tolerance zone* or an ideal feature that represents, for instance, a virtual condition (Fig. 6.4). Datums allow tolerance zones to be located or orientated and virtual conditions to be defined (for example, the *maximum material virtual condition*, according to ISO 2692).

6.1.1   Association of Datums

A datum is a theoretically exact point, axis, or plane that is obtained through an association criterion with the datum feature, that is, an ideal feature (plane, point or line) which is associated with the true (extracted and filtered) surface. The objective of such an association is to obtain an *unambiguous identification* of unique datums or datum systems.

In order to establish an associated feature, it is necessary to first perform a partition, then an extraction, a filtration and, finally, an association. The filtration should retain the highest points of the real integral feature.

For example, in the case of a datum feature made up of a flat surface of a part, the extracted surface of a true geometry is obtained, through a partition and extraction process, to which an ideal tangent plane of a perfect geometrical form is associated (situation feature, Fig. 6.5). In practice, the ideal tangent plane is approximated from the smooth surface of a matching plane of a control system, as shown in Fig. 6.6.

The associated features, used to establish the datums, simulate contact with the real integral features in a way that ensures that the associated feature is outside the material of the non-ideal feature. When the result of this process is ambiguous, then the associated feature must minimise the maximum distance (normal to the associated feature) between the associated feature and the filtered feature (Fig. 6.7).

The association process adopted to obtain an axis datum, starting from a true feature (integral) is shown in Fig. 6.8. The extracted feature is obtained through an extraction and filtration process; the derived feature (axis of the associated cylinder), and therefore the datum, are thus obtained through an association process with an ideal cylinder. In the same way as for a planar surface, in the case of cylindrical elements, the associated feature should minimise the maximum distance between the associated feature and the filtered feature (Fig. 6.9).

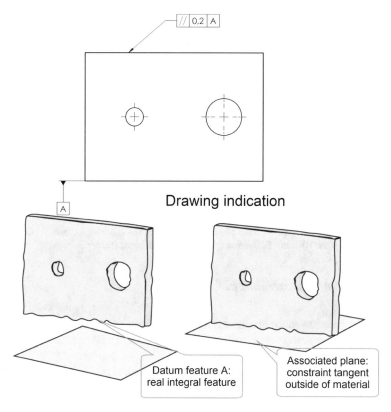

Fig. 6.5 The association process used to obtain a datum: the extracted surface is obtained from the true geometry, through a partition and extraction process, and an ideal tangent plane is associated with such a surface

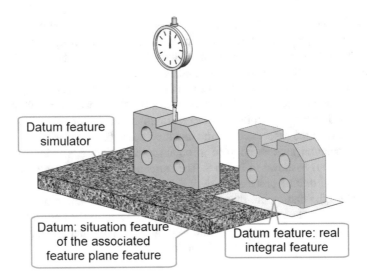

Fig. 6.6 If the datum feature is the flat surface of a workpiece, the associated feature is obtained from a theoretical envelope plane (e.g. a tangent plane) of perfect geometrical form, and it may be approximated from the matching granite plane of the control device

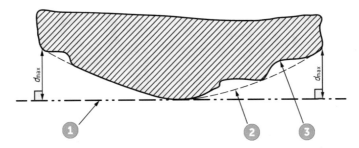

Fig. 6.7 The association method for a planar surface using the *minimax* criteria. 1. Outside the material tangent plane (datum). 2. Filtered feature. 3. Real integral feature

6.1.2 Datum: Mathematical Concepts

Datums and datum systems are theoretically exact geometric features used together with implicit or explicit theoretically exact dimensions (TED) to locate or orientate the tolerance zones and virtual conditions.

A datum consists *of a set of situation features* for an ideal feature (feature of perfect form) **associated** with the identified datum features of a workpiece. Since these ideal features have the task *of blocking the degrees of freedom of the tolerance zones*, the geometrical type of these associated features should be one of the **invariance classes** reported in Table 6.1.

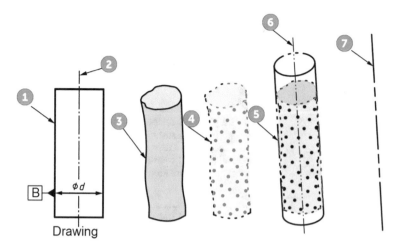

Fig. 6.8 The association and derivation process used to obtain a datum axis from a cylinder: 1. Nominal integral feature. 2. Nominal derived feature. 3. Real integral surface of the workpiece (datum feature). 4. Extracted integral feature; 5. Association with an ideal cylinder. 6. Derived feature (axis of the associated cylinder). 7. Datum (*situation feature*)

Fig. 6.9 The association method for a cylindrical surface using the *minimax* criteria. 1. Real integral feature. 2. Filtered feature. 3. Maximum inscribed cylinder

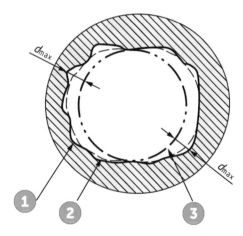

Each feature has 6 degrees of freedom (translations along the x, y, z axes and rotations around the x, y, z axes). The invariance class corresponds to the degrees of freedom (which remain and are not locked). Such a class describes the displacement of the feature (translation, rotation) for which the feature *is kept identical in space*.

In Table 6.1, all the surfaces are classified into seven classes on the basis of the degrees of freedom for which the surface is invariant. Situation features (point, straight line, plane) are defined for each class of surface.

When a single surface or a collection surface is identified as a datum feature, the invariance degrees for which the surface is invariant should be identified and

Table 6.1 All surfaces are classified into seven classes on the basis of the degrees of freedom for which the surface is invariant

Invariance class	Unconstrained degrees of freedom	Situation features
Spherical (i.e. sphere)	3 rotations around a point	Point
Planar (i.e. plane)	1 rotation and 2 translations	Plane
Cylindrical (i.e. cylinder)	1 translation and 1 rotation	Straight line
Helical	1 translation and 1 rotation	Straight line
Prismatic (i.e. prism)	1 translation	Plane, straight line
Revolute (i.e. cone)	1 rotation	Straight line, point
Complex (i.e. Bezier)	None	Plane, straight line, point

compared with Table 6.1 in order to determine the set of situation features (point, straight line, plane, or a combination thereof) that constitutes the datum.

6.2 Indication of Datum Features in Technical Documentation

A datum feature is identified with a capital letter, written within a square, and connected to a triangle that is placed on the feature itself (Fig. 6.10). It is suggested to avoid the use of the letters I, O, Q, X, Y and Z, which could give rise to interpretation problems, while, in the case of complex drawings, it is possible to make use of double letters (AA, BB, etc.).

In the same way as for the tolerance indications, a triangle, with its identification letter, can be located:

Fig. 6.10 The three ways of indicating a flat surface as a datum feature

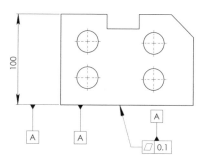

(a)   on the contour line of the feature surface or on an extension line of the feature outline (but clearly separated from the dimension line) if the datum feature is the surface itself (Fig. 6.11); it is also possible in this case to use a leader line that points directly towards the datum surface (Fig. 6.12).

(b)   on an extension of the dimension line of a feature of size when the datum is the feature axis or centerplane of the thus dimensioned part (Fig. 6.13); in this case, for space reasons, it is possible to substitute one of the arrowheads of the measurement line with a datum triangle symbol.

It is necessary to pay particular attention to the correct indication of a datum applied to a feature of size on a drawing: the line of the symbol should coincide with the dimension line of the datum feature, and the indication errors shown in Fig. 6.14 should be avoided.

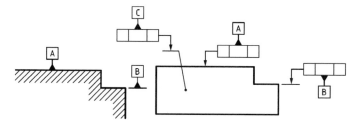

Fig. 6.11 Indications of a datum feature. A triangle does not need to be filled in

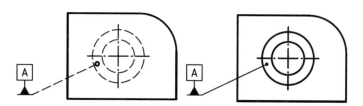

Fig. 6.12 Indications of a datum feature: the dashed leader line indicates that the datum is on the other side (when the considered surface is hidden)

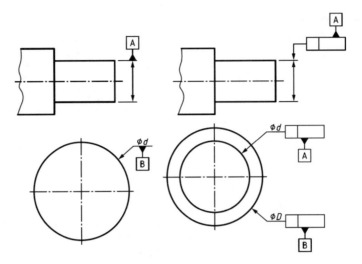

Fig. 6.13 Indication of a datum feature for a feature of size

Fig. 6.14 Possible errors in the indications of a datum axis. Cylindrical surfaces should not be used as datums. The symbol line should coincide with the dimension line of the datum feature

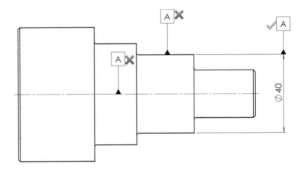

In light of all the definitions that have been given so far, does the datum symbol that is utilised in technical drawings indicate a datum or a datum feature (Fig. 6.15)? The symbol obviously indicates *a datum feature and not a datum*!

6.3 The Datum Systems in ASME Y14.5

The ASME Y14.5 standard makes a clear distinction between the concept of a **datum feature**, a datum and a **datum feature simulator**[1] A datum is an abstract geometrical feature (point, axis or plane from which a dimensional measurement is made), which

---

[1] The term "theoretical datum feature simulator" was replaced in the ASME Y14.5:2018 standard with the words "true geometric counterpart".

Fig. 6.15 Does the symbol, that is used in technical drawings, indicate a datum or a datum feature? The symbol obviously indicates a datum feature and not a datum

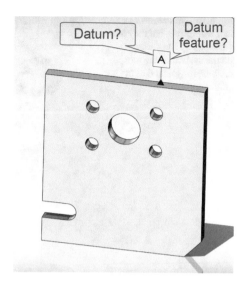

represents the perfect counterpart of a datum feature (e.g. an ideal plane or the axis of the perfect geometrical counterpart).

Simulated datums are conceptually perfect (almost physically perfect), and they represent a bridge between the imperfect real world of datum features and the perfect imaginary world of datums. Ultimately, it is opportune to distinguish between the real datum feature of a workpiece (named *datum feature*) and the datum, that is, the equivalent theoretical datum (plane, axis or centerplane), as simulated by the associated inspection or manufacturing equipment.

The datum system of a tool machine is shown in Fig. 6.16: the production equipment has the task of aligning the features of the workpiece with the datums of the machine (for example, datum feature A is aligned with the clamping machine and datum feature B is made to coincide with the z axis). In short, no datums exist on a workpiece, but they are simulated by the datum feature system of the tool machine.

The **true geometric counterpart** is the theoretically perfect boundary used to establish a datum from a specified datum feature. True geometric counterparts have a perfect form, and a basic orientation and location relative to other true geometric counterparts for all the datum references in a feature control frame. Furthermore, the true geometric counterparts are adjustable in size, when the datum feature applies an RMB (**Regardless Material Boundary**), and are fixed at the designated size, when an MMB (**Maximum Material Boundary**) or an LMB (**Least Material Boundary**) is specified.

The relationship between the primary datum feature and its true geometric counterpart **constrains** the degrees of freedom. Table 6.2 shows some examples of degrees of freedom constrained by primary datum features, RMB, in the same way as for the concept of invariance class.

Fig. 6.16 No datums exist on a workpiece, but they are simulated by the datum feature system of the tool machine (physical datum simulators)

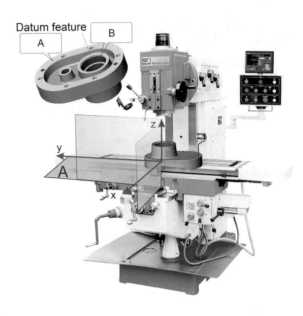

Table 6.2 Constrained degrees of freedom for primary datum features

Feature type	True geometric counterpart	Unconstrained degrees of freedom	Datum
Spherical	Sphere	3 rotations around a point	Point
Planar	Plane	1 rotation and 2 translations	Plane
Cylindrical	Cylinder	1 translation and 1 rotation	Axis
Width	Two opposed planes	1 rotation and 2 translations	Centerplane
Linear extruded shape	Extruded shape	1 translation	Centerplane, axis
Conical	Cone	1 rotation	Axis, point
Complex	Complex shape	None	Axis, centerplane, point

6.3.1 Establishing a Datum

In the case of a datum surface made up of a flat surface of a component, the datum
is supplied by a theoretical envelope plane (e.g. a tangent plane, the true geometric
counterpart), of perfect geometrical form, and the simulated datum feature is constituted, for example, by the surface plate of the inspection device. As can be seen in
Fig. 6.17, it is also possible to define a theoretical envelope plane, of perfect form
(true geometric counterpart of a datum feature), on the surface plate; if the workpiece
is placed on the surface plate, the two theoretical planes might not coincide because
of the irregularities on the surfaces. In spite of this observation, a simulated datum is
in fact used as a datum in industrial practice. Such a distinction may appear excessive
but, in reality, many errors exist and much confusion arises in defining the datums
and in the inspection of the parts with measuring machines.

If, instead, the datum feature is a feature of size, the symbol should be associated
with a linear dimension (subject to tolerance) and should therefore be placed directly
on the dimension, as shown in Fig. 6.18; in this case, the datum is made up of the

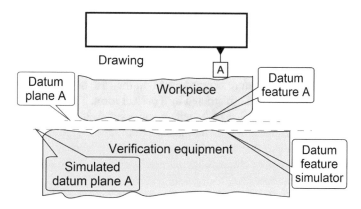

Fig. 6.17 Establishment of a datum plane

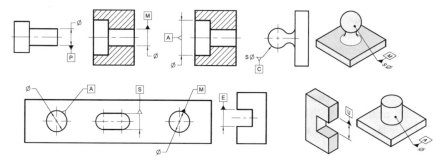

Fig. 6.18 Placement of Datum Feature Symbols on Features of Size

Fig. 6.19 When a
cylindrical feature is
designated as a datum
feature, the datum axis is
derived by placing the part in
an adjustable gauge
(self-centring chuck used as
a datum feature simulator)

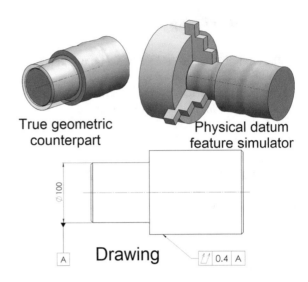

axis or centerplane, which is established by an ideal envelope surface.

Figure 6.19 shows the case of a datum made up of the axis of a cylinder; again in this case, it is possible to make a distinction between the axis of the smallest true geometric counterpart and the simulated datum feature, defined as the axis of the feature simulated by the verification gauge. As axes do not exist in the real world, the equivalent theoretical feature is represented, in the case of external features, by the axis of the smallest cylinder restricted by a perfect form. The datum is simulated by the verification device, for example, with a gauge or a self-centring device.

In the case of cylindrical holes, the simulated datum can be established through the use of a cylindrical gauge, with the maximum inscribable diameter, or an expandable chuck; whatever the type of chuck, once inserted into the hole, it does not take on a fixed shape, and it is necessary to regulate it so that any displacements are of the same magnitude in all directions (it is obvious that no modifiers should be applied to the datum).

In short, it is necessary to pay a great deal of attention to the characterisation of the simulated theoretical cylinder (which the ASME standards define as **Actual Mating Envelope**, AME[2]). In the simplest case, the cylinder is oriented according to the axis of the imperfect hole, but when the axis of the hole is defined on the basis of one or two datums, the simulated cylinder (**Related Actual Mating Envelope**) is arranged according to the DRF or *Datum Reference Frame*.

Figure 6.20 further clarifies the meaning of the *related* and *unrelated* Actual Mating Envelopes: datum A is a plane, while datum B is the axis of a 34 mm cylinder, oriented with an error of 0.05 mm with respect to plane A; the 10 mm hole is located with respect to datums A and B. The axis of the feature of size is

---

[2]A similar perfect feature counterpart of the smallest size that can be contracted about an external feature or of the largest size that can be expanded within an internal feature so that it coincides with the surface at the highest points.

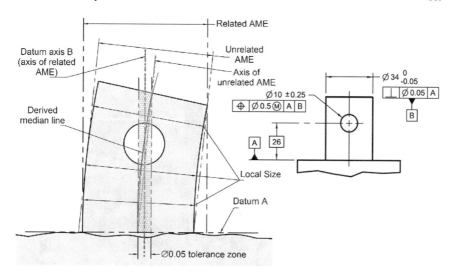

Fig. 6.20 Related and Unrelated AME in the ASME standard

obtained from the smallest restricted orientated cylinder (the *unrelated actual mating envelope*) and it is useful to verify a perpendicularity error with respect to datum A. The smallest circumscribed cylinder perpendicular to A (the *related actual mating envelope*) is used to determine the axis of datum B. The derived centerplane is a curve made up of the central points of the transversal section perpendicular to the axis of the feature of size, and it is used to determine the straightness error.

It is necessary to avoid confusing the axis of a feature of size (obtained through the smallest circumscribed cylinder orientated according to the feature, Fig. 6.21) for the axis of the datum B (perfect geometrical counterpart perpendicular to datum A).

6.3.2 Location of a Workpiece in a Datum Reference Frame

Simulated datums have the purpose of defining a DRF (*Datum Reference Frame*), that is, the datum system of 3 perpendicular planes that define the origin for the measurements and which allow the workpiece to be blocked during an inspection (Fig. 6.22). In short, the indications of a DRF system supply an immediate description of the orientation and of the location of a workpiece (and of the tolerance zones), thus **making the inspection operations univocal and repeatable**.

In order to limit the movement of a part so that repeatable measurements can be made, it is necessary to restrict the six degrees of freedom. In fact, each piece that has to be controlled or worked in a datum system has six degrees of freedom (3 linear and 3 rotational, Fig. 6.23). It is possible to show that, in order to eliminate the 6 degrees of freedom, it is necessary to block the workpiece in a datum reference

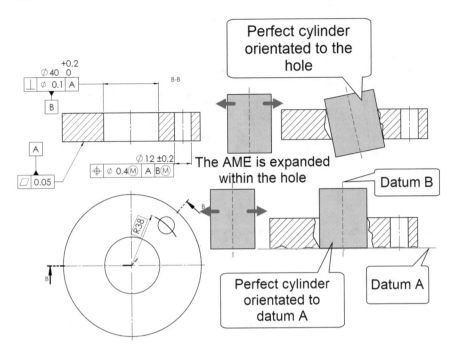

Fig. 6.21 It is necessary to be careful not to confuse the axis of a feature size (obtained through the largest circumscribed cylinder orientated according to the feature), which is useful to control the perpendicularity of a 40 mm hole, with the axis of datum B (axis of the largest inscribable cylinder perpendicular to datum A)

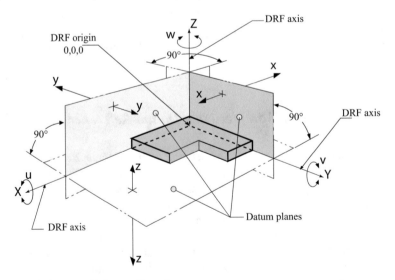

Fig. 6.22 DRF (Datum Reference Frame)

Fig. 6.23 The six degrees of freedom of a workpiece: any movement in the space can be attributed to three possible translations in the direction of the datum axes X, Y and Z and to three rotations, u, v and w, around the same axes

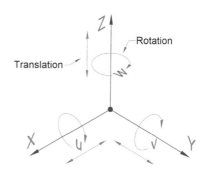

frame with 3 perpendicular planes, named the primary, secondary and tertiary planes, respectively.

Let us consider the drawing of the component shown in Fig. 6.24, whose holes are located with respect to the three datums A, B and C; in order to conduct a measurement test, datum feature A is put into contact with datum feature simulator A, thus establishing a primary datum with at least three points of contact, and eliminating a degree of linear freedom Z and two rotational degrees of freedom, that is, u and v (Fig. 6.25).

Datum feature B is then put in contact with datum feature simulator B, and in this way secondary datum B is established, with a minimum of two points of contact and elimination of a linear degree, X, and a rotational one w (Fig. 6.26). Finally, tertiary datum C is defined with just a single point of contact and the last degree of freedom Y is eliminated by putting datum feature C in contact with its simulator (Fig. 6.27).

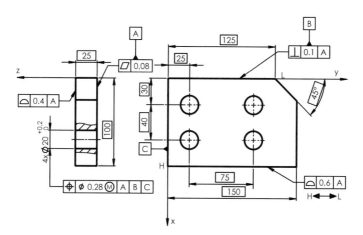

Fig. 6.24 In order to make the identification of a datum reference frame clearer, the ASME Y14.5 standard introduced the possibility of identifying the axes of the datum system on a drawing in order to offer an immediate description of the orientation and location of a workpiece (and of the relative tolerance zone), thus making the measurement test operations univocal and repeatable

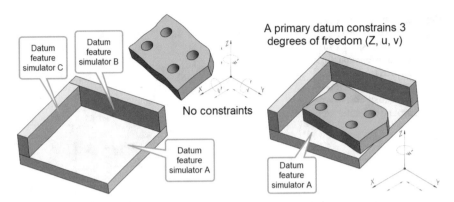

Fig. 6.25 Datum feature A of the component is put in contact with the datum feature simulator, and in this way a primary datum is established, with a minimum of three contact points and elimination of a degree of freedom Z and two rotational ones, that is, u and v. The X, Y and w degrees of freedom still have to be constrained

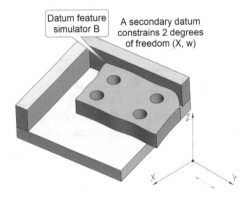

Fig. 6.26 Subsequently, datum feature B is put in contact with datum feature simulator B, and in this way the secondary datum is established, with a minimum of two contact points and elimination of a linear degree, X, and a rotational one w. The translation along Y still has to be constrained

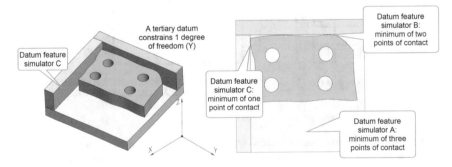

Fig. 6.27 Final blocking of the workpiece: tertiary datum C is defined with at least one point of contact (and elimination of the last degree of linear freedom Y), thus datum feature C is placed in relation to its simulator

Fig. 6.28 The order of the datums in the tolerance frame indicates the sequence of the inspection operations; as can be seen, the location dimension of the hole changes according to the datum plane on which the workpiece is placed

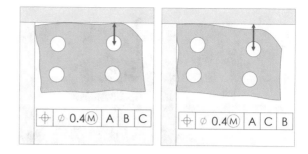

It is necessary to pay particular attention to the order of the datum sequence as it influences the result of an inspection. In fact, if the inspection procedure of the workpiece in Fig. 6.28 is considered, it can be noted that, as the location tolerances have been indicated in the datum order A, B and C in the frame, the inspection procedure should be conducted after locating the workpiece in the datum system according to the exact same order. A change in this order (i.e. A, C, B) would influence the measurement test, as can be seen in the same figure.

6.3.3   Selection of Datum Features

The starting point for a correct dimensioning is the identification of a datum reference frame with 3 perpendicular planes (DRF), and in this way functioning and mating of the parts are guaranteed. As previously mentioned, it is erroneous to think that the datums should only be established in the manufacturing design phase on the basis of the requirements of the various operations. Datums should in fact be selected on the basis of the functional requirements of a part in order to allow the functional relationships to be communicated on the drawing. The datums should instead always be established on the basis of the functional requirements because, if the datums are chosen on the basis of the technological location of the workpieces, the tolerance available for manufacturing may be reduced.

The choice of the datum features should obviously be considered in function of the assembly requirements of the parts, and their sequence very often reproduces the logical assembly sequence. The case shown in Fig. 6.29 can be considered as an example: the part with 2 holes should be mated to the plate by means of two fixed fasteners. The logical sequence that allows an appropriate choice of the three datums to be made is:

1. identification of the feature that **orientates** the workpiece in the assembly, in order to establish the primary datum; in Fig. 6.30, the feature that orientates the workpiece is constituted by mating surface A.
2. Identification of the feature that **locates** the workpiece in the assembly, in order to establish the secondary datum; datum feature B that locates the workpiece is represented by the axis of the two cylindrical features.

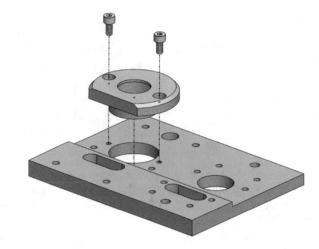

Fig. 6.29 Choice of the functional datums for a mating with fixed fasteners

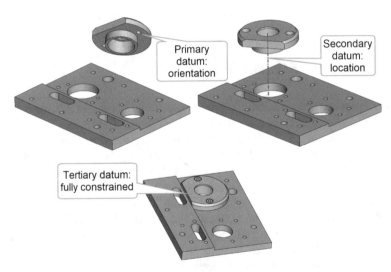

Fig. 6.30 The feature that is used to orientate the workpiece is made up of mating surface A (which is defined as the primary datum), while the feature that is considered to locate it is the axis of the two cylindrical features (which is defined as a secondary datum). Finally, the tertiary datum that blocks the workpiece, which is constituted by the milled surface of the component and of the corresponding surface of the plate, is added

3. Identification of the feature that **blocks** the workpiece in the assembly, in order to establish the tertiary datum. Finally, tertiary datum C that blocks the workpiece, which is constituted by the milled surface of the component and of the corresponding surface of the plate, is added.

4. **Qualification** of the datum features through the application, orientation and location of opportune form tolerances.
5. **Location** of all the geometrical features with reference to the datum features, while utilising the profile tolerances to locate the surfaces.

The complete dimensioning of the upper component is reported in the ASME drawing in Fig. 6.31. The dimensioning of the lower plate is obtained by following the same previously outlined rules.

However, it is not always necessary to use three datums. The case shown in Fig. 6.32 can be considered as an example: the cover with the holes should be mated with the tank through fixed fasteners; the feature that orientates the workpiece is constituted by mating surface A (which is defined as the primary datum), while the feature that locates it is represented by the axis of the cylindrical part (which is defined as the secondary datum). When the axis of a cylinder is used as the datum feature, it is necessary to imagine using two perpendicular planes, whose intersection determines the datum axis, and which can rotate freely around the same axis (Fig. 6.33).

If the assembly of the cover takes place as in Fig. 6.34, that is, the tongue must be mated with the relative groove, three datums are required, that is, it is necessary

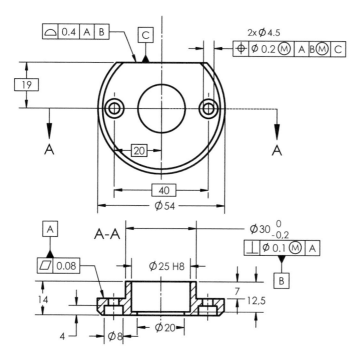

Fig. 6.31 Complete dimensioning of a component with the 3 established functional datums. The primary datum is qualified with a planarity tolerance and the secondary one with a perpendicularity tolerance. Finally, the two holes in the plate are located with a location tolerance with respect to the three established datums

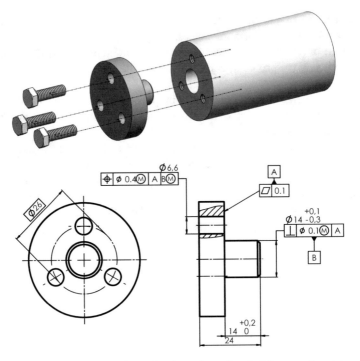

Fig. 6.32 Selection of the primary datum (used to orientate) and of the secondary datum (used to locate)

Fig. 6.33 The axis of a cylinder that is used as a datum feature is associated with two imaginary perpendicular planes, whose intersection determines the datum axis, which can rotate freely around the same axis

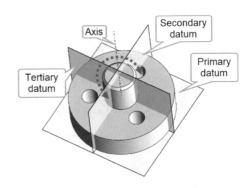

to add the tertiary datum that blocks the workpiece and which is constituted by the centerplane of the groove.

In this case, the two imaginary planes can no longer rotate and are constrained in a precise location (Fig. 6.35). Table 6.3 summarises the main differences between the ISO and ASME standards pertaining to the specification of a datum.

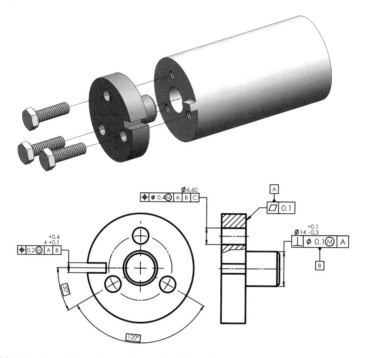

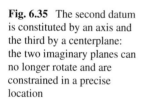

Fig. 6.34 Selection of the primary datum (used to orientate), of the secondary datum (used to locate) and of the tertiary datum (used to block)

Fig. 6.35 The second datum is constituted by an axis and the third by a centerplane: the two imaginary planes can no longer rotate and are constrained in a precise location

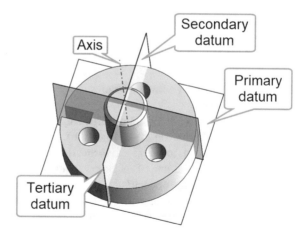

6.4 Examples of Datum Systems in the ISO Standards

In an ISO standard, a datum system is a set of two or three situation features established in a specific order from two or more datum features. A **primary datum** is a datum that is not influenced by any constraints from other datums. A **secondary**

Table 6.3 The main differences between the ISO and ASME standards pertaining to the specification of a datum

Datum system	Theoretical constraints	Establishing a datum	Size of the datum boundary	Physical datum
Situation features of the collection of the associated features (i.e. plane, point and straight line)	Invariance class	Association with an ideal feature	MMVC, LMVC	–
DRF (Datum reference frame with three mutually perpendicular datum planes)	Degree of freedom	True geometric counterpart	MMB, LMB	Simulated datum

datum is a datum, in a datum system, that is influenced by a primary datum orientation constraint in the datum system. A **tertiary datum** is a datum, in a datum system, that is influenced by constraints from the primary datum and the secondary datum in the datum system. A datum system is constituted by an *ordered* sequence of two or three datums. This order defines the orientation constraints that should be followed for the association operation: the primary datum imposes orientation constraints on the secondary datum and tertiary datum; the secondary datum imposes orientation constraints on the tertiary datum.

6.4.1 Datum System with Three Single Datums

6.4.1.1 Three Perpendicular Planes

The datum system is composed of three single datums with orientation constraints (perpendicularity) between them (Fig. 6.36). The datums are used together sequentially, in a given order, to orient and locate the tolerance zone relative to a plane, one of its straight lines and one of its points (equivalent to three planes). These are the situation features of the collection of the three associated planes.

The primary datum is an associated plane; the secondary datum is an associated plane that respects the orientation constraint of the primary single datum; the tertiary datum is an associated plane that first respects the orientation constraint from the primary datum and then the one from the secondary datum (Fig. 6.37). The situation features are a *plane* (corresponding to the primary datum), a *straight line* (the intersection between this plane and the plane corresponding to the secondary datum) and a *point* (the intersection between the straight line of the secondary datum and the plane corresponding to the tertiary datum).

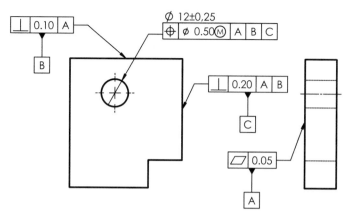

Fig. 6.36 Datum system with three perpendicular planes

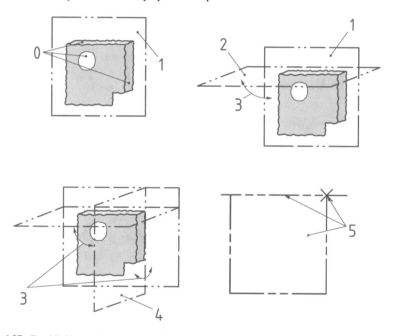

Fig. 6.37 Establishing a datum system from three perpendicular planes: 0. Datum feature: real integral features. 1. The plane associated with the primary datum feature identified by datum letter A. 2. The associated plane (secondary datum) identified by datum letter B, with an orientation constraint from primary datum. 3. The perpendicularity constraint. 4. The associated plane (tertiary datum) identified by datum letter C, with orientation constraints from the primary datum and secondary datum. 5. The datum system: plane (primary datum), straight line (intersection between the primary and secondary datums) and point (intersection of the three datums)

6.4.1.2   A Plane and Two Perpendicular Cylinders

In this case, a planar surface and two perpendicular cylinders are used to establish a datum system by considering a variable size of the cylinders and a perpendicularity constraint between the axis of the associated cylinders and the associated plane (Fig. 6.38).

The datum system is characterised by the situation feature of the collection of a plane and two perpendicular cylinders, with a constraint location between them. The primary datum in the datum system is an associated plane; the secondary datum is an associated cylinder, which respects the orientation constraint from the primary single datum; the third associated feature is an associated cylinder with a perpendicularity constraint from the primary datum and a location constraint from the secondary datum (Fig. 6.39). The situation features are a plane (first associated feature), a point (intersection between the plane and the axis of the second associated feature) and a straight line (intersection between the associated plane and the plane containing the two axes, Fig. 6.40).

6.5   Type of Datum

6.5.1   Centerplanes (Width-Type Features of Size)

The design intent of the drawing shown in Fig. 6.41 is to achieve symmetry of the median plane of the internal slot with respect to the median plane of the external element. In this case, the real integral surface resulting from the collection of two nominally parallel planar surfaces, which is a feature of size, is used to establish a datum by considering the size variable.

The two real integral surfaces of the workpiece indicated in Fig. 6.42 are obtained after partition/extraction/collection and constitute a feature of size of variable size.

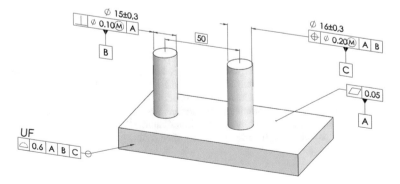

Fig. 6.38 Datum system taken from a plane and two cylinders

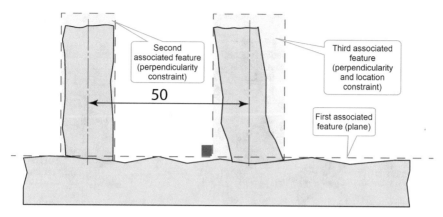

Fig. 6.39 The primary datum in the datum system is an associated plane; the secondary datum is an associated cylinder that respects the orientation constraint from the primary single datum; the third associated feature is an associated cylinder with a perpendicularity constraint from the primary datum and a location constraint from the secondary datum

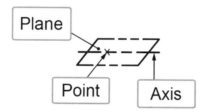

Fig. 6.40 The situation features are a plane (first associated feature), a point (intersection between the plane and the axis of the second associated feature) and a straight line (intersection between the associated plane and the plane containing the two axes

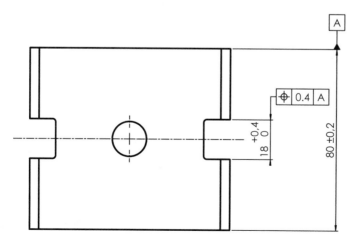

Fig. 6.41 The design intent is the symmetry of the median plane of the internal slot with respect to the median plane of the external element

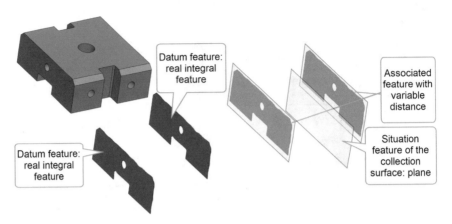

Fig. 6.42 Establishing a single datum from a feature of size (two parallel opposite planes)

The single datum is characterised by the situation feature obtained from the collection of two parallel planes associated with the real integral features used to establish the single datum.

The association is made with an internal parallelism orientation constraint (the distance between the two planes is variable). The invariance class of the collection of nominal surfaces is *planar* and the situation feature is a plane (the median plane of the two associated planes).

In the ASME standard, the central datum plane of an external feature is the symmetry plane between two parallel planes which, at the minimum distance, are in contact with the corresponding surfaces of the workpiece (Fig. 6.43); vice versus, the central datum plane for internal features is constituted by a symmetry plane between two parallel planes which, at the maximum distance, are in contact with the corresponding surfaces of the workpiece (Fig. 6.45). Figure 6.44 shows a simulation

Fig. 6.43 ASME standard: the central datum plane of an external feature is the symmetry plane between two parallel planes which, at the minimum distance, are in contact with the corresponding surfaces of the workpiece

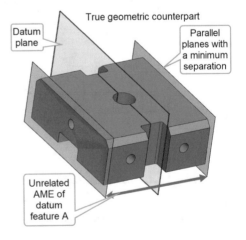

Fig. 6.44 Physical datum feature simulator in ASME: Datum centerplane of a component. The simulation of the datum is obtained with the median plane of the two parallel surfaces of the vise clamps

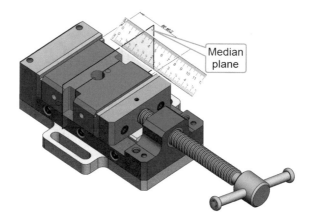

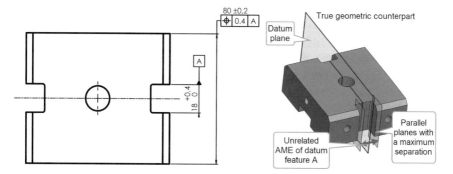

Fig. 6.45 The central datum plane for internal features is constituted by a symmetry plane between two parallel planes which, at the maximum distance, are in contact with the corresponding surfaces of the workpiece

(according to the ASME standards) of the centerplane assumed as the datum of the component shown in Fig. 6.41.

It is necessary to pay particular attention to the location of the datum feature symbol (Fig. 6.46). Placing the symbol on the extension of the dimensioning line indicates a feature of size, that is, the centerplane of the feature, otherwise, the indicated datum feature is the lateral plane.

6.5.2 Common Datum

In some cases, the datum is established by utilising two or more datum features in a simultaneous way until a common datum is defined, that is, a datum established from two or more datum features after simultaneous associations without any specific order, but with interrelated constraints.

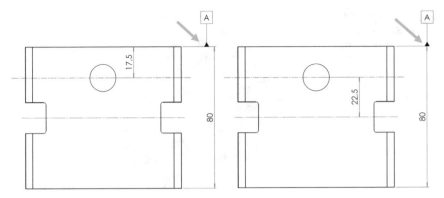

Fig. 6.46 It is necessary to pay particular attention to the location of the datum feature symbol. Placing the symbol on the extension of the dimensioning line indicates the centerplane of the feature, otherwise the indicated datum feature is the lateral plane

The design intent of the drawing shown in Fig. 6.47 is to simultaneously use two integral, nominally cylindrical and coaxial surfaces, which are features of size, to establish a datum, by considering their sizes variable and the orientation constraint (parallelism) and location constraint (coaxiality). The common datum (axis) is used to orient and locate the tolerance zone of the run-out control. This axis is the situation feature of the collection of the two associated cylinders.

The common datum in Fig. 6.48 is characterised by the situation feature of the collection of two coaxial cylinders associated with the real integral features used to establish the common datum. The association is made with internal constraints: zero distance and parallel (coaxial). The invariance class of the collection of nominal surfaces is cylindrical, and the situation feature is the common axis of the two cylinders.

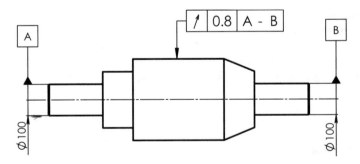

Fig. 6.47 A common datum is established simultaneously with two integral cylindrical and coaxial surfaces. The common datum is used to orient and locate the tolerance zone of the run-out control. It is important to underline that the two letters in the frame indicate that the datum features are separated by a hyphen as they refer to a single datum

Fig. 6.48 Establishing a common datum from two coaxial cylinders

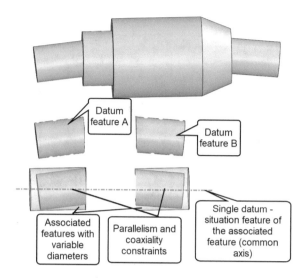

The common datum axis A-B in the ASME standard is the axis of the true geometric counterpart (smallest pair of coaxial circumscribed cylinders). The physical datum simulator of the A-B axis is a couple of coaxial centring devices (Fig. 6.49).

The design intent of the drawing shown in Fig. 6.50 is the simultaneous use of two integral, nominally planar surfaces constrained to be coplanar, which are not features of size, to establish a datum. The datum is used to orient the parallelism tolerance zone relative to this plane (the situation feature of the collection of the two associated planes).

In this case the real integral surfaces are obtained after partition/extraction and collection, and the common datum is characterised by the situation feature of the collection of two coplanar planes associated with the real integral features used to establish the common datum (Fig. 6.51). The association is made with coplanar internal constraints (i.e. zero distance and parallel). The invariance class of the collection of nominal surfaces is planar and the situation feature is a plane.

Fig. 6.49 Physical datum feature simulator of the common datum axis as in the ASME standard. The simulation of the datum is obtained with two coaxial centring devices

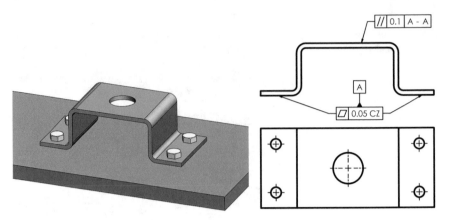

Fig. 6.50 An example of two coplanar surfaces used to establish a common datum A-A

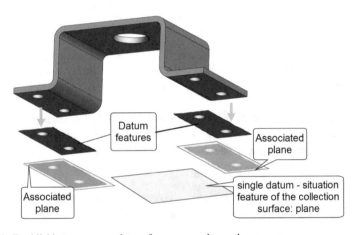

Fig. 6.51 Establishing a common datum from two coplanar planes

6.5.2.1   Pattern of Holes as a Common Datum

A pattern of holes can be chosen as a common datum to simultaneously establish a datum, taking into consideration the size, orientation, and location of the associated cylinders. Two parallel holes are used as a common datum in Fig. 6.52. In this case, the two integral cylindrical and parallel surfaces, which are features of size, are used simultaneously to establish a datum by considering the size of the cylinders variable and the orientation constraint (parallelism) and location constraint (76 mm) between the two axes. The situation feature is made up of a plane that passes through the two axes of the holes and the median straight line of the axes of the two associated cylinders (Fig. 6.53).

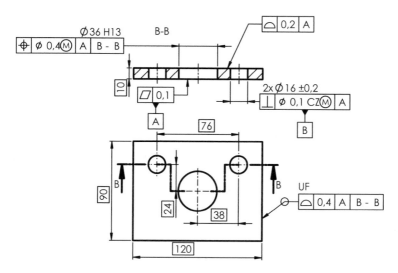

Fig. 6.52 The two 16 mm holes are used as a common datum. It should be noted that, in this case, two datums are sufficient to eliminate all the degrees of freedom, as the two "hole axis" datums are constrained as far as the orientation (parallelism) and location are concerned

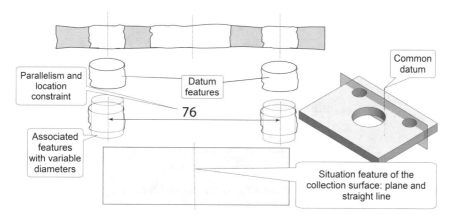

Fig. 6.53 Establishing a common datum from two parallel cylinders: the situation features are the plane containing the two axes and the median straight line of the axes of the two associated cylinders

The same component is shown in Fig. 6.54, where the common datum is indicated with a **DV** (*distance variable*) modifier, which allows only the orientation (parallelism), and not the distance between the two holes, to be constrained.

The design intent of the drawing in Fig. 6.55 is to simultaneously use four integral, nominally identical and cylindrical surfaces, which are features of size, together with parallel axes to establish a datum by considering the size of cylinders variable and the orientation constraints (parallelism) and location constraints (76 × 48 mm).

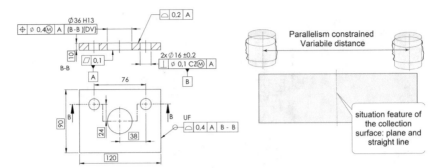

Fig. 6.54 The common datum is indicated with a DV (distance variable) modifier. The two 16 mm holes simultaneously establish a common datum, but the axes are only constrained in orientation (parallelism) and not in location. The verification is obviously conducted with a different piece of equipment from the previous case

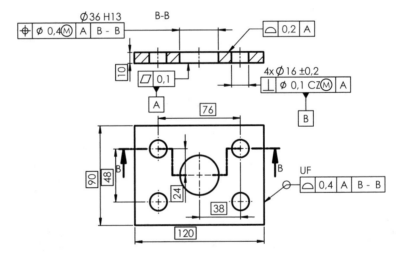

Fig. 6.55 A pattern of 4 holes can be utilised as a common datum, by associating cylinders of variable diameter, in order to establish a datum reference frame. The datum axes are constrained in orientation (parallelism) and location by the theoretically exact dimensions (76 and 48 mm). It should be noted that, again in this case, two datums are sufficient to eliminate all the degrees of freedom

The datum axes in Fig. 6.56, are constrained, as far as the orientation (parallelism) and location are concerned, by the theoretically exact dimensions of the distances (76 and 48 mm) and a couple of identical capital letters are used for indication purpose

Four cylinders that are simultaneously constrained in orientation and location are associated with the true integral features. As in the previous case, the final common

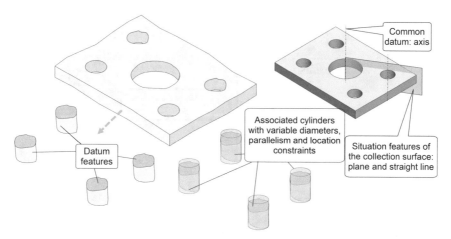

Fig. 6.56 Four cylinders that are simultaneously constrained in orientation and location are associated with the true integral features. As in the previous case, the final common datum is defined by the central axis of the pattern and by a plane that passes through the axis of one of the associated cylinders

datum is defined from the central axis of the pattern and from a plane that passes through the axis of one of the associated cylinders.

6.5.3 Conical Surfaces as Datum Features

When a cone is defined as a datum feature, the resulting datum is made up of a line (axis) and a point (vertex of the cone, as in Fig. 6.57, or a particular point along the axis, which is defined by the position at which the diameter of the section is specified, as in Fig. 6.58). An ideal cone, with the same angular opening, is associated with the real integral surface of the cone, whose axis defines the common datum which, together with the point, eliminates 5 degrees of freedom (Fig. 6.59).

The design intent of the drawing shown in the Fig. 6.60 is to use the integral, nominally conical surface, which is a feature of size, to establish a datum by considering its size fixed. Only the datum is used to orient and locate the tolerance zone relative to an axis (coaxiality). In this case, this axis is the situation feature of the associated cone, and the point *is not involved* in the location of the tolerance zone.

Figure 6.61 shows an ASME drawing in which conical primary datum feature A constrains five degrees of freedom, including translation in Z. In this case, the YZ and ZX planes may rotate because rotation w is not constrained. The datum reference frame used to locate the 10 mm diameter hole originates at the apex of the conical true geometric counterpart. When orthographic views are used, the rectangular coordinate axes should be labelled in at least two views on the drawing.

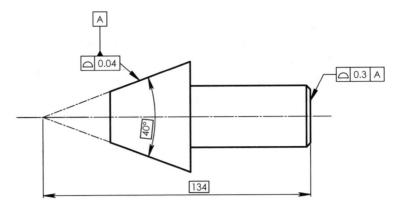

Fig. 6.57 The design intent of the drawing is to use the integral, nominally conical surface, which is a feature of size, to establish a datum by considering its size fixed. Only the datum is used to orient and locate the profile tolerance zone relative to its situation features, which are the axis of the associated cone and a point along this axis

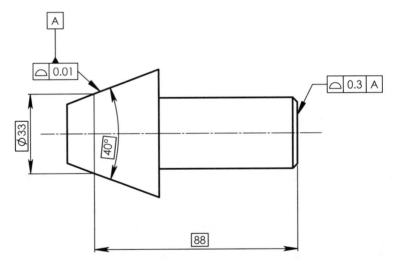

Fig. 6.58 The design intent is the same as is in the previous figure, but this time the situation features are the axis of the associated cone and a particular point along the axis, defined by the location at which the section diameter is specified

6.6   Datum Features Referenced at MMR and LMR (Size Datum)

As the dimension of a feature of size may vary from the maximum to the least material, when it is used as a datum feature, it is necessary to specify whether

Fig. 6.59 The real integral surface is obtained after partition/extraction. The datum is characterised by the situation features of the cone associated with the real integral feature without external constraints. The invariance class of the nominal surface is revolute and the situation features are the axis of the cone and a particular point along this axis (vertex or defined by the location)

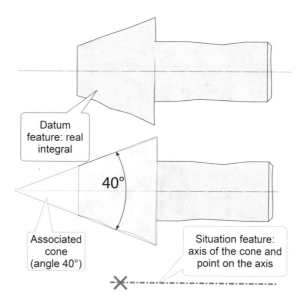

Fig. 6.60 Only the datum is used to orient and locate the tolerance zone relative to an axis (coaxiality). In this case, this axis is the situation feature of the associated cone, and the point is not involved in the location of the tolerance zone

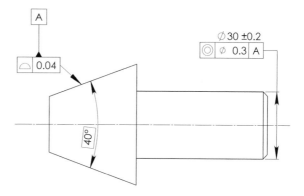

it is used for the maximum material conditions (defined as MMVC[3] (Maximum Material Virtual Condition), LMVC[4] (Least Material Virtual Condition), or in any intermediate conditions[5] (default)).

When there are no modifiers, default conditions are assumed, and in this case, as has been observed so far, it is possible to derive the datum from the associated perfect geometry (axis or centerplane). On the other hand, if there is a material modifier, the datum is the axis or centerplane of the virtual conditions, and therefore has a fixed size.

---

[3] ASME Maximum Material Boundary, MMB.

[4] ASME Least Material Boundary, LMB.

[5] ASME Regardless of the Material Boundary, RMB.

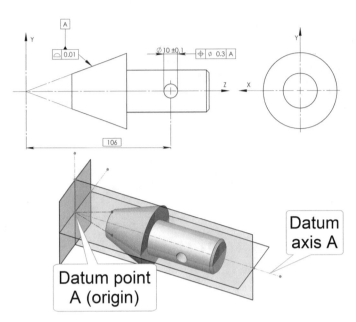

Fig. 6.61 ASME drawing: conical datum feature referenced to constrain five degrees of freedom. When orthographic views are used, the rectangular coordinate axes should be labelled in at least two views on the drawing

The case of the dimensioning of the plate in Fig. 6.62 should be considered: the 16 mm central hole is referenced with respect to a datum reference frame constituted by three planes, that is, A, B and C, and in turn constitutes a datum feature D (Diameter 16 mm) of the two 8 mm holes that are located with respect to the A, D and C datum system.

Since a maximum material modifier appears next to the indications of datum feature D, the extracted feature of the datum feature should not violate its MMVCs, which is a cylinder an MMVS diameter:

$$MMVS = MMS - 0, 3 = 16 - 0, 3 = 15, 7 \, mm$$

The axis of the MMVC virtual condition is theoretically exactly oriented (perpendicular) to datum A, and in the theoretically exact location (28 mm) relative to datums B and C (Fig. 6.63).

Figure 6.64 shows the case of an external datum feature B (Ø28 mm), qualified by means of perpendicularity with respect to primary datum A. The extracted feature of the datum feature B should not violate the maximum material virtual condition

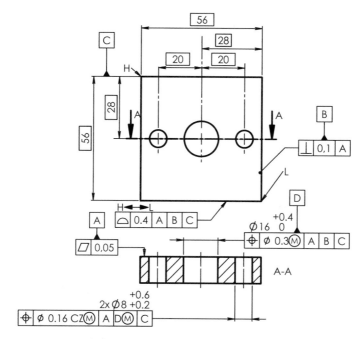

Fig. 6.62 The 16 mm central hole is referenced with respect to an A, B and C datum system, which in turn constitutes a datum feature D of the two 8 mm holes that are located with respect to the A, D and C datum system. A maximum material modifier appears next to the indication of datum feature D

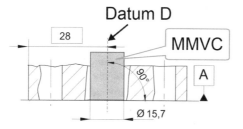

Fig. 6.63 The extracted feature of datum feature D should not violate its MMVC (Ø15.7 mm), The axis of the MMVC virtual condition is theoretically exactly oriented (perpendicular) to datum A, and in the theoretically exact location (28 mm) relative to datums B and C

MMVC (Ø28,2 mm), which is obtained by summing the geometrical tolerance with the maximum material dimension, MMS. At the same time, the extracted feature of the toleranced internal feature should not violate the MMVC (MMVS Ø9,6 mm) oriented with respect to datum A and located at 0 mm from the axis of the MMVC of datum feature B.

Apart from being applied to axes, a maximum material modifier or a least material modifier can also be applied to centerplanes. In fact, in the case of the drawing in

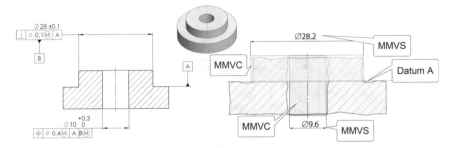

Fig. 6.64 The control of the central hole takes place through an orientation with respect to datum A and coaxially with respect to datum B. The extracted feature of the toleranced internal feature should not violate the MMVC (MMVS Ø9,6 mm) and the extracted feature of datum feature B should not violate the MMVC (MMVS Ø28,2 mm)

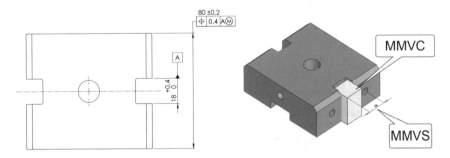

Fig. 6.65 By placing the maximum material modifier next to the indications of datum feature A, the datum becomes the median plane of a block under MMVC conditions (in this case, coinciding with the maximum material dimensions of the slot, that is, 18 mm)

Fig. 6.65, by indicating the maximum material modifier next to datum feature A, the datum becomes the centerplane of a block under MMVC conditions (in this case, coinciding with the maximum material dimensions of the slot, that is, 18 mm). The extracted feature of the slot should not violate the MMVC.

Analogously, in the case of external features (Fig. 6.66), the datum becomes the centerplane of two symmetrical blocks, at the distance from the MMVC conditions, that should not be violated (in this case, coinciding with the maximum material dimensions of the datum, that is, 80.2 mm).

It is possible to apply the maximum, or the least material modifiers, to a common datum constituted by a pattern of holes, as shown in Fig. 6.67. In this case, the virtual condition is defined by 4 cylinders perpendicular to datum A, constrained in orientation and spacing and of constant dimensions (Ø15,8 - 0,1 = Ø15,7 mm). The extracted features of the four tolerance features should not violate the MMVC, which has a diameter MMVS = Ø15,7. Moreover, the four datum features are theoretically exactly oriented to datum A and in the theoretically exact location relative to each

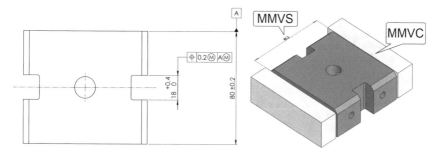

Fig. 6.66 In the case of an external datum (features of size) indicated at MMVC, the datum becomes the median plane of two symmetrical blocks, at a distance from the virtual conditions (in this case, coinciding with the maximum material dimensions of the datum, that is, 80.2 mm)

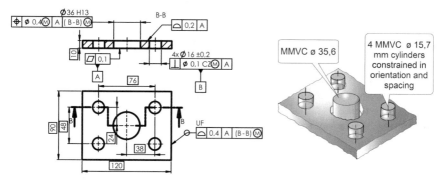

Fig. 6.67 In the case of a common datum constituted by a pattern of holes, it is possible to utilise the maximum and least material modifiers. In this case, the virtual condition is defined by 4 cylinders perpendicular to datum A, constrained in orientation and spacing and of constant dimensions (15,8-0,1 = 15,7 mm)

other (76 × 48 mm). The extracted feature of the central Ø16 mm hole should not violate the MMVC, which has a Ø35,6 mm diameter.

6.7   Locked or Released Degrees of Freedom from a Datum

When a datum system is used in a geometrical specification, the situation feature of the tolerance zone is oriented or located on the basis of the datum system defined in the datum section of the tolerance indicator. However, it is possible to vary the number of degrees of freedom eliminated from the datums, according to the design requirements, by utilising PL, SL, PT, >< modifiers after the indication of the datum and enclosing them in square brackets (Fig. 6.68). The complementary indications

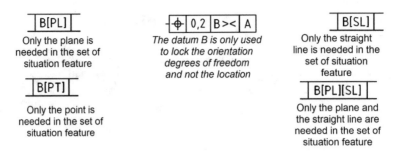

Fig. 6.68 By default, a datum restricts all the possible degrees of freedom but, through some modifiers, it is possible to indicate the degrees of freedom that are blocked by a datum feature

[PL], [SL] and [PT] are only used when the situation feature of a plane, a straight line or a point is needed, respectively.

The complementary indication $><$ is only used to lock the orientation degrees of freedom, and not the location, and should be omitted when the geometrical characteristic only controls the orientation of the feature (e.g. a perpendicular specification).

In Fig. 6.69, for example, the task of locating the tolerance zone has passed from datum B to datum A. Datum B only controls the orientation (that is, it only restricts two degrees of freedom instead of three).

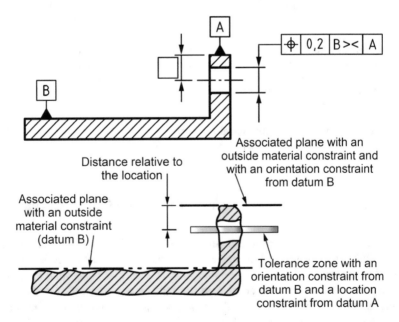

Fig. 6.69 Example of a datum with the $><$ modifier. Datum B should orientate and locate the tolerance zone (constrained by three degrees of freedom). The use of the "$><$" modifier only allows the orientation to be controlled, while the location of the tolerance zone is left to datum A

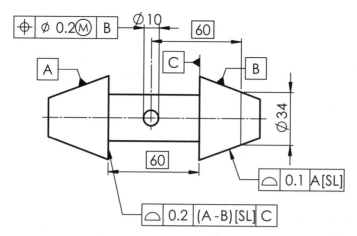

Fig. 6.70 The situation feature used to establish datum A is the axis of the cone. The situation features used to establish datum B are the axis of the cone and one point on the axis. When a complementary indication (SL, PL or PT) applies to all the elements of the collection of surfaces of a common datum, the sequence of letters that identify the common datum should be indicated within parentheses

Only datum A in Fig. 6.70 is used to orient and locate the tolerance zone relative to one of the situation features of the associated cone (only the axis of the associated cone). Datum B is used to orient and locate the tolerance zone of the 10 mm hole to the situation features of the associated cone, that is, the axis of the cone and a particular point along this axis, defined by the location. When a complementary indication (SL, PL or PT) applies to all the elements of the collection of surfaces of a common datum, the sequence of letters that identify the common datum should be indicated within parentheses.

6.7.1   Customised Datum Reference Frame in the ASME Standards

Datums in the ASME standard have the task of defining the **DRF** (*Datum Reference Frame*), that is, the 3 perpendicular plane datum system that defines the origin for the measurements and allows a workpiece to be blocked during an inspection or during working operations. A primary datum eliminates three degrees of freedom (2 rotational, that is, u and v, and one linear, z). A secondary datum eliminates two degrees of freedom, (linear y and rotational w). Finally, a tertiary datum eliminates the last degree of freedom, that is, of translation x. In short, a datum system is defined to restrict some degrees of freedom related to its use.

The ASME Y14.5 standard introduced the possibility of customising the number of degrees of freedom eliminated from each plane of the DRF. For example, it is

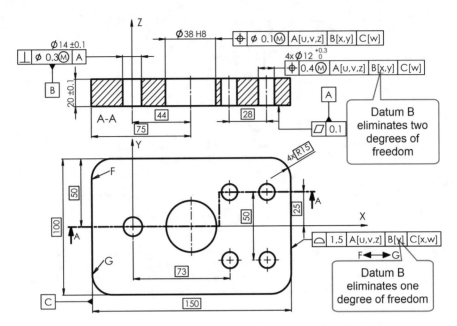

Fig. 6.71 In the ASME standards, the eliminated degrees of freedom are indicated in square brackets. It is possible to vary the number of degrees of freedom eliminated by each datum by, for example, transferring the task of limiting the degrees of freedom along direction x from datum B to datum C

possible to indicate, in brackets, the number and type of degrees of freedom that have been eliminated, with reference to the datum system. Primary datum A in Fig. 6.71 eliminates three degrees of freedom, that is, u, v and z; secondary datum B (axis of a 14 mm hole) eliminates two degrees of translational freedom, that is, x and y, and tertiary datum C has the task of restricting rotation w in order to control the location error of the 38 mm hole and of the four 14 mm diameter holes.

Instead, the task of limiting the degree of freedom along direction x has passed from datum B to datum C to locate and orientate the profile tolerance zone, and datum B therefore only eliminates one degree of freedom.

When multiple datum reference frames exist, and it is desirable to label the axes of the coordinate system (X, Y, and Z), any labelled axes should include a reference to the associated datum reference frame. In Fig. 6.72, the X, Y and Z axes for the three datum reference frames are identified as [A, B, C], [A, B, D] and [A, B, E], and they represent the datum features of each DRF.

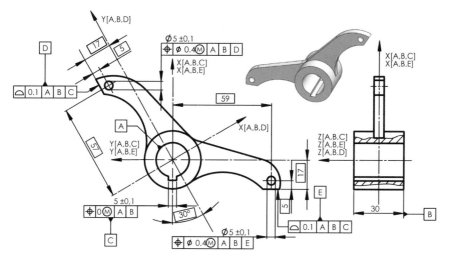

Fig. 6.72 When multiple datum reference frames exist, the labelled axes should include a reference to the associated datum reference frame. The X, Y and Z axes of the three datum reference frames are identified as [A, B, C], [A, B, D] and [A, B, E], and they represent the datum features of each DRF

6.8   Datum Targets

When it is preferred not to use a complete integral feature to establish a datum feature, it is possible to indicate portions of the single feature (areas, lines or points) and their dimensions and locations. These portions are called **datum targets**. They usually simulate the interface between the considered single feature of the workpiece and one or more contacting ideal features (assembly interface features or fixture features).

Datum targets are used in the case of complicated and irregular forms, such as those produced for moulding or casting, or for non-planar and distorted surfaces. Datum targets are a compromise between the functioning of a feature and repeatability of the measurement and, even when indicated on the drawing of a workpiece, they in fact describe the form and location of the control features that are utilised to simulate a datum plane.

Datum targets should be used in the following cases:

(a) when only a portion of a part feature is functional and can be used as a datum feature;
(b) when an irregular or regular form prevents the use of a planar surfaces as a datum;
(c) when the workpiece becomes unstable during the verification process, once it has been located with its datum surface.

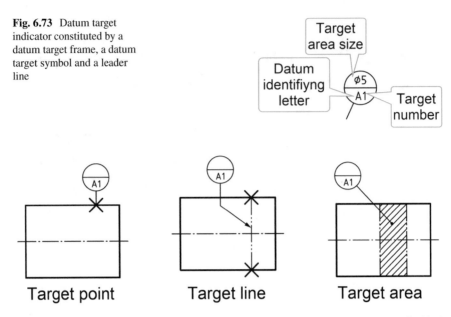

Fig. 6.73 Datum target indicator constituted by a datum target frame, a datum target symbol and a leader line

Fig. 6.74 Datum targets are made up of points, lines and planar areas (of any form) or cylindrical contact planes, bordered by a 5.1 line, according to ISO 128-4

A datum target is indicated by a **datum target indicator**, which is constituted by a datum target frame, a datum target symbol and a leader line. Datum targets are indicated by a circle divided into two compartments by a horizontal line (Fig. 6.73). The lower compartment is reserved for a letter, which represents the datum feature, and for a number, which indicates the number of the datum target. The upper compartment is reserved for complementary information, such as the dimensions of the datum target zone. If there is not enough space in the compartment, the information may be placed outside the circle and joined with a leader line.

Datum targets are established (see Fig. 6.74) with points, lines and planar contact areas (of any form) or cylindrical areas, bordered by a 5.1 line according to ISO 128-4. Each datum target should be located by means of theoretically exact dimensions, in that they do not refer to the features of a workpiece, but instead define the size and characteristics of the control system. Moreover, the theoretically exact dimensions not only ensure repeatability of the measurements, but can also refer to the degree of precision of the inspection system.

If the datum target on a workpiece is a point, it is indicated with a cross; in the case shown in Fig. 6.75, the primary datum is defined by at least three points of contact (A1, A2 and A3), the secondary datum by two points (B1 and B2) and the third by point C1; the control equipment, whose pins simulate the datum targets in the drawing, is also visible in the same figure.

If the datum target is made up of a line, it is indicated with two crosses united by a fine 5.1 double-dashed line which, when this line is not closed (open line), is

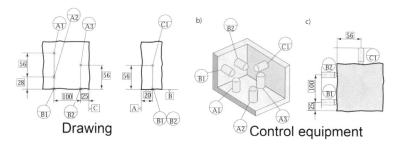

Fig. 6.75 Datum targets made up of points and the relative control equipment with spherical-tipped gauge pin verification

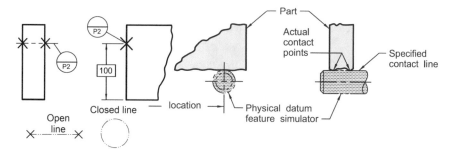

Fig. 6.76 Datum targets made up of lines simulated in control equipment by the side of a gauge pin

terminated by two crosses[6], as shown in Fig. 6.76; finally, if the datum target is made up of a flat contact zone, it is indicated with a hatched area whose borders are defined by a fine 5.1 double-dashed line (Fig. 6.77) or even by a cross, when there is no risk of misunderstanding in the identification of the datum.

If the datum is on the opposite side to that indicated by the symbol, the leader line is dashed, as shown in Fig. 6.78.

In order to express the use of datum targets, the datum feature identifier should be repeated close to the datum feature indicator, and should be followed by a list of numbers (each number separated by a comma followed by a space) to identify the datum target identifier (Fig. 6.79).

It is also possible to specify datum targets on cylindrical surfaces, as shown in Fig. 6.80 [1]: the A1 datum target is specified with a circular datum target line (which is visible as a line in the image), while datum B is specified with a cylindrical datum area; the verification requires the use of variable diameter gauges.

Figure 6.81 shows a typical application of a datum target: the assembly requirement of the plated component requires the indication of an "area" datum target.

The drawing of a workpiece, obtained by means of casting, is shown in Fig. 6.82,

---

[6]Crosses are not used in this case in the ASME standard.

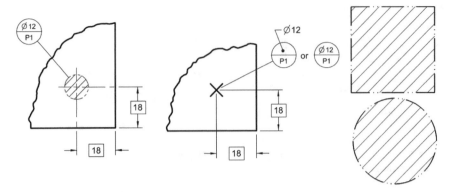

Fig. 6.77 If the datum target is made up of a planar contact zone, it is indicated by a hatched area in which the borders are defined by a fine 5.1 double-dashed line

Fig. 6.78 If the datum target is on the opposite side to that indicated by the symbol, the leader line is dashed

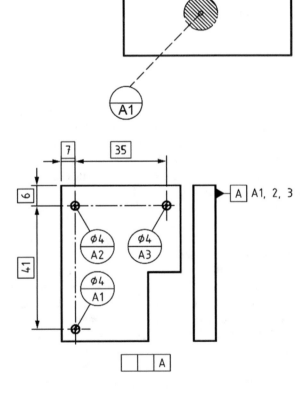

Fig. 6.79 To express the use of datum targets, the datum feature identifier should be repeated close to the datum feature indicator, and should be followed by a list of numbers (each number separated by a comma followed by a space) to identify the datum target identifier

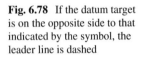

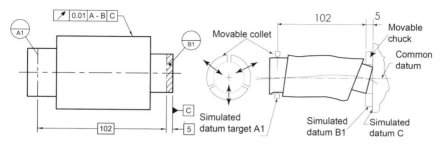

Fig. 6.80 Cylindrical datum target area and line with the simulated gauge

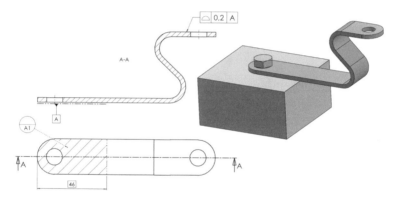

Fig. 6.81 Assembly of the components requires the use of a datum target area. A dashed line is used to indicate the involved surface

with indications of three datum targets, which define a datum system with three perpendicular planes, in which:

– the three datum targets A1, A2 and A3 specify primary datum A;
– the two datum targets B1 and B2 specify secondary datum B;
– the datum target C1 (point) specifies tertiary datum C.

The control equipment is visible in Fig. 6.83; datum plane A is simulated by means of 3 pins with a diameter of 40 mm, datum plane B is simulated by the straight lines between the two pins and datum C is defined by the conical tip of the equipment. Each datum target is located by means of a theoretically exact dimension to indicate that it is necessary to apply the same degree of precision as the control system. Therefore, all the dimensions that locate the datum targets refer to the datum target simulators of the control gauge, and not to the datum features of the workpiece.

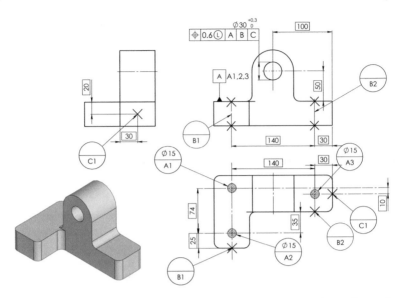

Fig. 6.82 Application of datum targets to establish a datum system. A primary datum plane is established by at least three target points. A secondary datum plane is usually established by two targets. A tertiary datum plane is usually established by one target. A combination of target points, lines and areas may be used and datum targets are located by means of theoretically exact dimensions

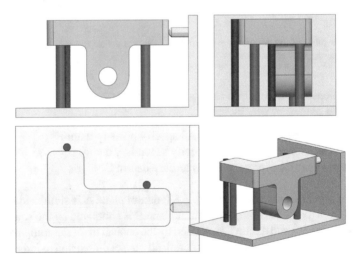

Fig. 6.83 Control equipment for a workpiece that is in contact with the true geometric counterparts for datum features. The datum targets defined by points are simulated by pins with spherical heads. The datum targets defined by lines are simulated by the straight lines of a cylindrical pin. Finally, the datum targets of areas are defined by a pin with a flat head

6.8.1 Contacting Feature

As has already been seen, a datum in the ISO 5459 standard is an "ideal feature which is fitted to the datum feature with a specific association criterion". In other words, reference is made to an "associated feature" to establish a datum, for example, by associating an ideal tangent plane to a true surface. An interesting novelty introduced by the standard is the definition of "contacting feature", which defines an ideal feature that is different from a datum feature indicated on the drawing and is associated to such a drawing by means of a "contact" operation (Fig. 6.84).

Let us consider, for example, the drawing of the component shown in Fig. 6.85, produced for moulding, in which datum feature A is the support plane, the second

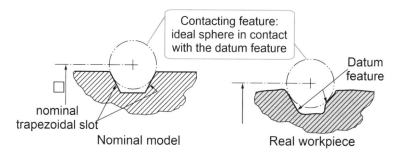

Fig. 6.84 A contacting feature is an ideal feature, with a theoretically exact geometry, which is different from the nominal geometry of the integral geometrical feature with which it is in contact

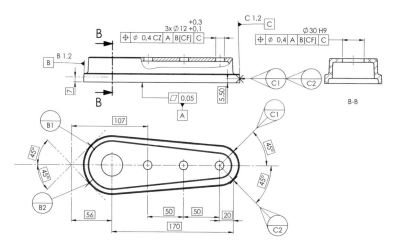

Fig. 6.85 The CF symbol, placed next to the datum feature B indications, defines an ideal feature (V block), which is different from the feature indicated on the drawing and associated to it with a "contact" operation. The two moveable datum target modifiers C1 and C2 are used to define the direction in which the location of the datum target is not fixed

datum is made up of two datum targets, that is, B1 and B2, while the symbols of the two moveable target datums, C1 and C2, are used for the third datum.

Datum targets B1 and B2 are defined by the interface between the cylindrical surface of the workpiece and a *contacting feature*. The distance between datum targets B1 and B2 is variable and depends on the actual diameter of the cylinder and the contacting feature, which is defined, in this case, by a "V" block of angle 90° (the associated feature used to establish the datum). The CF symbol, placed next to the datum feature B indications, defines this ideal feature (V block), which is different from the feature indicated on the drawing and associated to it with a "contact" operation.

The modifier [CF] implies that some portions of the workpiece are used to establish the datum, and that the location of the contact between the contacting feature and the workpiece cannot be determined exactly (as it depends on the dimensions and the geometry of the real workpiece).

The two moveable datum target modifiers C1 and C2 are used to define the direction in which the location of the datum target is not fixed. The direction of the motion is indicated by the direction given by the moveable modifier and not by the leader line.

The intention of the design is to allow a simulated datum target (Moveable Datum Target Simulator) to translate or move along a specified direction. By default, a Moveable Datum Target Simulator can translate perpendicularly to the contact surface, but another translation direction may be specified in the drawing and indicated with the theoretically exact dimensions. In our case, the movement direction is at 45° with respect to the median plane of the workpiece.

The workpiece is controlled by means of the functional gauge shown in Fig. 6.86, in which the B1 and B2 fixed datum targets are made up of the sides of the V block, while the C1 and C2 datum targets are simulated by two moveable pins. Figure 6.87 illustrates the control procedure, with the two retractable C1 and C2 pins, which fix the workpiece in the gauge. C1 and C2 move synchronously [2].

Another example is shown in Fig. 6.88, where a perfect distinction is made between the indications of the datum feature (truncated conic surface) and the

Fig. 6.86 Functional gauge for the control of a workpiece: the fixed datum targets B1 and B2 are constituted by the sides of the V block, while the C1 and C2 datum targets are simulated by two moveable pins. C1 and C2 move synchronously

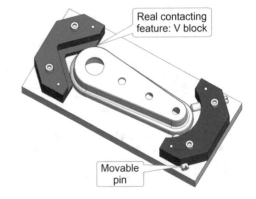

Real contacting feature: V block

Movable pin

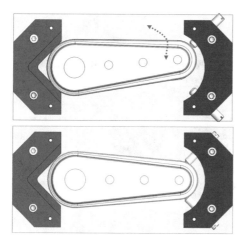

Fig. 6.87 Control procedure, with two retractable pins which fix the workpiece to the gauge. The part is staged on the gauge and datum target simulators C1 and C2 are engaged

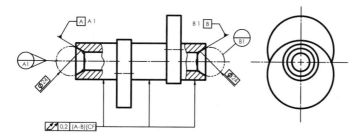

Fig. 6.88 A perfect distinction between the indication of the datum feature (truncated conic surface) and contacting features A1 and A2, simulated by two 24 mm spheres, is obtained by means of the indications of contact feature CF. Explicit reference to the datum target, through the A1 and B1 indications next to the A and B labels, should also be noted

contacting features A1 and B1, simulated by two 24 mm diameter spheres. Explicit reference to the datum targets, through indications A1 and B1 next to the labels A and B, can also be noted. The corresponding control procedure is carried out through two 24 mm spheres of which one corresponds to moveable datum A1. The contacting feature (the sphere) is a different feature from the datum feature indicated on the drawing, which corresponds to the internal conical surface of the component under examination (Fig. 6.89).

Figure 6.90 shows the association process that should be adopted to obtain datum axis A: the datum is derived from an associated cylinder (e.g. the smallest theoretically circumscribed cylinder) in order to locate the tolerance zone of the profile. The same component is verified with a different setup in the case in which the CF contacting feature symbol is used (Fig. 6.91). In this case, two tangent cylinders, with dimensions, orientation and location fixed on the drawing, are used.

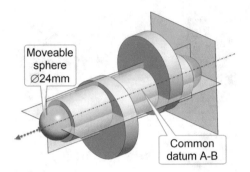

Fig. 6.89 A control procedure carried out by means of two 24 mm spheres, one of which corresponds to movable datum target A1

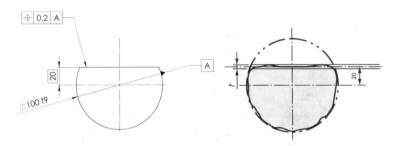

Fig. 6.90 Association of a single datum without a modifier [CF]: datum A is derived from an associated cylinder (e.g. the smallest theoretically circumscribed cylinder) in order to locate the tolerance zone of the profile

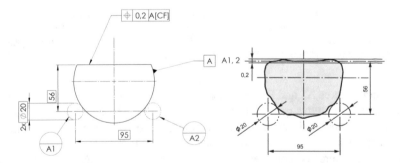

Fig. 6.91 Association of a single datum with a modifier [CF]: the same component as that shown in the previous figure is verified with a different setup. In this case, two tangent cylinders, with the dimensions, orientation and location fixed on the drawing, are used

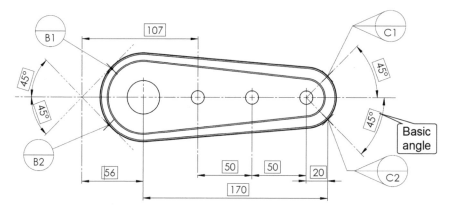

Fig. 6.92 The "movable datum target" symbol in ASME standards is the same as in the ISO standards, but the movement may be indicated through the addition of a line that indicates the direction

6.8.2 Datum Targets in the ASME Standards

The symbols of the datum targets in the ASME Y14.5 standard are basically identical to those of the ISO standard, albeit with some slight differences:

1. The datum target line is the same as in the ISO standard (a long-dashed double-dotted line, type 5.1 as in ISO 128-24), but the line does not terminate with two crosses.
2. The "movable datum target" symbol is the same as in the ISO standard, but the movement may be indicated for orthographic views through the addition of a line that indicates the direction of movement (see Fig. 6.92). The line element should be specified with one or several basic angles[7].
3. Alternatively, the direction of movement, for drawings that include X, Y, and Z axes to represent the datum reference frame(s), may be indicated using a unit vector designation consisting of i, j, k components (corresponding to the X, Y, and Z axes of the coordinate system), placed in brackets and adjacent to the "movable datum target" symbol. The vector direction is towards the surface of each datum feature. See Fig. 6.93.
4. When using datum features defined by datum targets in a feature control frame established by fewer than three mutually perpendicular planes, the datums that are the basis of the datum reference frame should be referenced (Fig. 6.94). The targets that provide a definition of the datums referenced in the feature control frame should be specified in a note, such as: *where only datum feature A is referenced, datum features B and C are invoked only to relate the targets that establish datum A.*

Some useful advice on datum targets is given hereafter:

---

[7]In ISO 5459, the direction is only given by the moveable modifier, and not by the leader line.

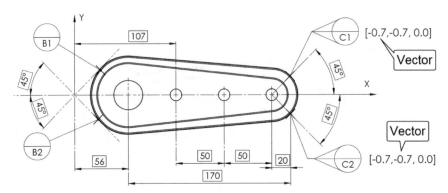

Fig. 6.93 The movement direction may be indicated using a unit vector designation, consisting of i, j, k components, placed in brackets

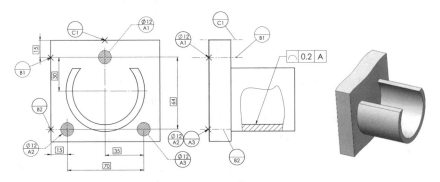

Fig. 6.94 If primary plane A is defined as a datum target, a secondary and a tertiary plane are only invoked to relate the targets that establish datum A

(1) the area indicated by a datum target in a drawing should NOT be realised or carved into the component;

(2) if the primary plane is defined as a datum target, both a secondary and a tertiary plane are mandatory;

(3) if the datum target is an area, contact with the setup feature does not necessarily have to take place over the whole area, as a series of any points within the area specified on the drawing is sufficient.

References

1. Krulikowski A (1999) Advanced concepts of GD&T, textbook based on ASME Y14.5 M—1994, Effective Training Inc.
2. Fischer BR (2009) The Journeyman's guide to geometric dimensioning and tolerancing: GD&T for the New Millennium. Advanced Dimensional Management Press

Chapter 7
Form Tolerances

Abstract It is possible to control four form tolerance types: straightness, flatness, roundness and cylindricity. The chapter covers all the concepts necessary to define specification operators (according to ISO 17450-2) and some procedures to establish the reference elements in order to define the deviation errors. Some terms related to form parameters are described such as peak-to-valley, peak-to-reference and reference-to-valley deviations. The ASME standards use the envelope requirements or Rule #1, according to which form tolerances are contained within the dimensional ones, and these tolerances are therefore only used with the purpose of limiting the error when the workpiece is produced with dimensions close to the least material condition.

7.1 Introduction

Form tolerances are used to establish the variation limits of a surface or of a feature of the ideal form indicated on a drawing; in short, the form error of a feature is limited with respect to its perfect and ideal counterpart (plane, line or circle).

It is possible to control four types of form tolerance: straightness, flatness, roundness[1] and cylindricity. The ASME standards use the envelope requirements or Rule #1, according to which form tolerances are covered by the dimensional ones, and these tolerances are therefore only used when necessary.

Since, as already mentioned, a profile tolerance controls not only form errors, but also size, location and orientation errors, a specific section is dedicated to this topic.

7.2 Straightness Tolerance

Straightness is the condition in which a linear feature (or any linear feature of a surface) results to be perfectly straight. Straightness is basically a characteristic of a line, such as an axis or an edge of a feature; however, this type of tolerance can also

---

[1] Circularity in the ASME standard.

© The Author(s), under exclusive license to Springer Nature Switzerland AG 2021
S. Tornincasa, *Technical Drawing for Product Design*, Springer Tracts
in Mechanical Engineering, https://doi.org/10.1007/978-3-030-60854-5_7

163

be applied to flat, cylindrical or conic surfaces, which are considered to be composed of an infinite number of longitudinal features.

A straightness symbol should obviously be placed in the drawing view where the feature explicitly appears straight. Figure 7.1 indicates the interpretation of a straightness tolerance of 0.03 mm placed on a cylindrical surface: each longitudinal line element of the cylinder should be between two parallel lines at a distance of 0.03, on a plane composed of the axis and the two lines themselves. A tolerance zone is bi-dimensional, and one of the two lines of the tolerance zone is orientated by the extreme points of the line element of a surface, while the other line is parallel to the first and distanced by the tolerance.

Three possible types of error are visible on the cylindrical surface shown in the same figure, that is, of concavity, convexity and of bending. A straightness tolerance applied to a flat surface only controls the straightness in the direction parallel to the projection plane.

In the case shown in Fig. 7.2, the straightness tolerance is limited by two parallel lines 0.1 mm apart, in the direction specified by the plane intersection symbol (or, in the case in which a symbol is missing, in the direction parallel to the projection plane). The tolerance zone is therefore limited by two parallel lines at a distance specified by tolerance t, at any intersection plane parallel to datum A (Fig. 7.3).

Figure 7.4 shows the application of a straightness tolerance of the surface of a shaft (integral feature). As the principle of independence comes into force by default, the following rules are valid [1]:

(1) A tolerance zone is applied to each line element individually.
(2) The maximum virtual boundary is obtained by summing the maximum material size and the straightness tolerance.

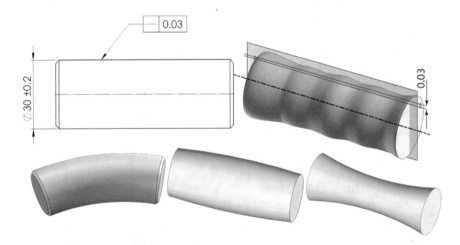

Fig. 7.1 Interpretation of a straightness tolerance of 0.03 mm placed on a cylindrical surface: each line element of the cylinder should be between two parallel lines 0.03 apart, in a plane composed of the axis and the two lines themselves

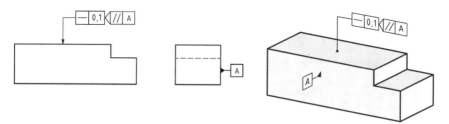

Fig. 7.2 Straightness tolerance applied to a flat surface: the direction is specified by the plane intersection symbol

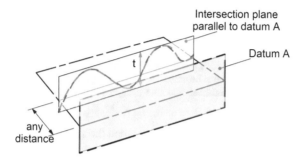

Fig. 7.3 Interpretation of the tolerance zone shown in Fig. 7.2. The tolerance zone is limited by two parallel lines at a distance specified by tolerance t, in any intersection plane parallel to datum A

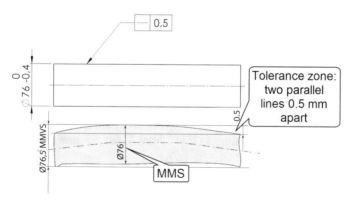

Fig. 7.4 Application of a straightness tolerance to the surface of a shaft. As the principle of independency comes into force by default, the maximum virtual boundary is obtained by summing the maximum material size with the straightness tolerance, which can be greater than the size tolerance

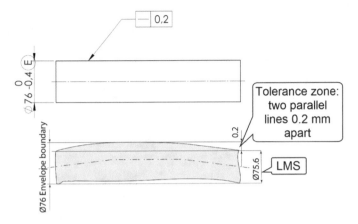

Fig. 7.5 In the case of the envelope requirement being applied (circled symbol E next to the diameter value), the straightness must be perfect when the component is produced at the maximum material and the geometrical tolerance should always be less than the size tolerance

(3) The straightness tolerance may be larger than the size tolerance.
(4) Each local size should fall within the size tolerance limits.

In the case of the application of the envelope requirement (circled E symbol), the straightness must be perfect when the component is produced at the maximum material condition (Fig. 7.5) and the geometrical tolerance should always be less than the size tolerance. It should be recalled that this case represents the default conditions in the ASME standards.

7.2.1 Straightness Parameters

The ISO 12780 standard explains the terms and concepts necessary to define specification operators (according to ISO 17450-2) for the straightness of integral features.

According to this standard, the extracted straightness line is a digital representation of the intersection of the real surface and a straightness plane, which includes the normal of the real surface (Fig. 7.6). The straightness profile is an extracted line intentionally modified by a filter. In the evaluation of a straightness deviation of an integral feature with a given tolerance, the straightness profile should be contained between two lines that are distant from each other by a value that is less than or equal to the specified tolerance value.

When determining the orientation of the tolerance zone, it is necessary to establish a reference line, that is, an associated line that fits the straightness profile according to specified conventions, and to which the deviations from straightness and the

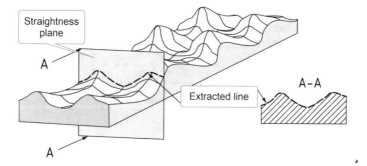

Fig. 7.6 The extracted straightness line is a digital representation of the intersection of the real surface and a straightness plane, normal to a real integral surface

straightness parameters refer. The ISO 12780-1 technical specification considers two procedures for the determination of the reference line:

- The minimum zone reference line method (MZ), which best satisfies the tolerance zone definition (Fig. 7.7).
- The least squares reference line method (LS), which provides a good approximation of the straightness deviation, although overestimating it, and is currently the most frequently used in coordinate measuring machines.

The local straightness deviation is defined as the deviation of a point on a straightness profile from the reference line, in the direction normal to the reference line.

Some terms related to straightness parameters are listed hereafter:

(1) **Peak-to-valley straightness deviation**, which is the value of the largest positive local straightness deviation added to the absolute value of the largest negative local straightness deviation. The GT modifier is used in specifications to indicate

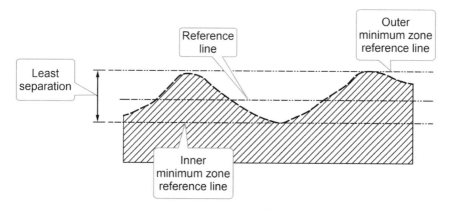

Fig. 7.7 Minimum zone reference line

Table 7.1 Deviation parameters. The peak height and the valley depth are only defined relative to the minimax (Chebyshev) association and the least squares (Gaussian) association

Symbol	Parameter
P	Reference-to-peak
V	Reference-to-valley
T	Peak-to-valley
Q	Root mean square (RMS)

that a form tolerance applies to the peak-to-valley deviation relative to the least squares reference element.

(2) **Peak-to-reference straightness deviation**, which is the value of the largest positive local straightness deviation from the least squares reference line.

(3) **Reference-to-valley straightness deviation**, which is the absolute value of the largest negative local straightness deviation from the least squares reference line.

The Table 7.1 shows the parameter specification elements that can be used for form specifications, i.e. specifications that do not reference datums. T may be used to indicate the total range of deviations, i.e. the default parameter. P, V should be used to indicate the peak height, and the valley depth, respectively. Q should be used to indicate the square root of the sum of the squares.

7.2.2 Straightness Tolerance Applied to a Feature of Size

Let us now examine the case in which the straightness tolerance is applied to a feature of size, that is, to a derived median line; in this case, the tolerance has been indicated on the dimension that expresses the diameter, and reference is therefore made to the median line which must remain within a three-dimensional tolerance area, limited by a cylinder with the same diameter as the tolerance itself, and this is indicated by placing the diameter symbol Ø before the number that expresses the tolerance value (Fig. 7.8). In this case, the straightness tolerance can be greater than the size tolerance of the diameter of the associated cylinder.

Figure 7.9 shows, for the case of a shaft, the concept of derived median line, obtained by means of a set of central points of the single sections. In practice, a cylinder, from which the axis is derived, is associated with the extracted surface (e.g. utilising a Gaussian interpolation). A line, to which a Gaussian circle is associated, and whose centre determines a point of the derived median line, is extracted from each section perpendicular to the axis.

When a geometrical control is applied to a feature of size, it is possible to use the maximum material requirement (MMR), with the advantages that are derived from the increase in the tolerances, as can be seen in Fig. 7.10. The combined effect of the size error and the straightness error generates a virtual size (MMVS), which represents the worst possible mating condition.

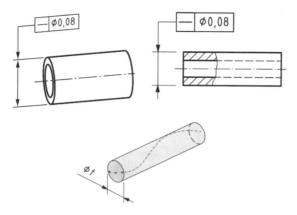

Fig. 7.8 A straightness tolerance is applied to a feature of size; in this case, the tolerance is indicated on the dimension that expresses the diameter, and reference is therefore made to the extracted median line of the cylinder; the tolerance zone is three-dimensional and limited by a cylinder with a diameter equal to the tolerance itself

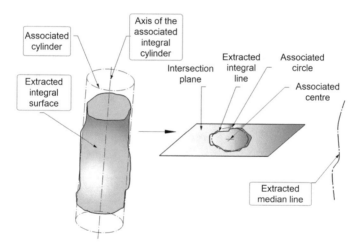

Fig. 7.9 The extracted median line obtained from a set of central points of the single sections perpendicular to the axis of the associated cylinder

The virtual condition (MMVC) is the configuration of the limit envelope of perfect form generated by the combined effect of the maximum material dimension and the geometrical tolerances. It should be noted that the virtual condition is always the one that corresponds to the worst possible mating conditions which, in the case of a shaft, are obtained by summing the value of the geometrical tolerance with the maximum diameter (that is, the maximum material size). In the case of a hole, the maximum material conditions are those that correspond to the minimum diameter, and the virtual condition MMVC is always represented by the worst mating conditions, that

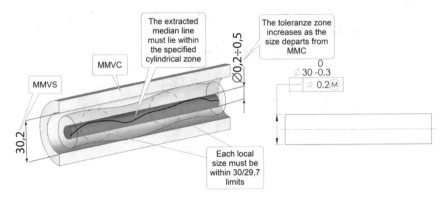

Fig. 7.10 When a straightness control is applied to a feature of size, it is possible to use the maximum material requirement. The combined effect of the size error and the straightness error generates a virtual size (MMVS), which represents the worst possible mating condition

is, those obtained by subtracting the geometrical tolerance value from the minimum diameter (Fig. 7.11).

It is also possible to combine a total straightness tolerance with a straightness tolerance on a specified length. In this case, a composite tolerance frame may be used, as in Fig. 7.12. The derived median line must remain within a 0.4 mm diameter cylinder along the entire length, but should not exceed 0,1 mm for each 25 mm of length.

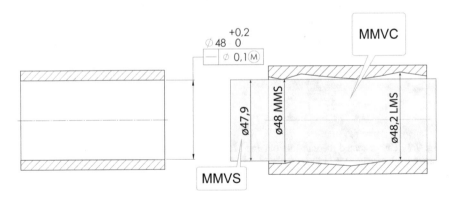

Fig. 7.11 In the case of a hole, the maximum material conditions are those that correspond to the minimum diameter (MMS), and the virtual condition MMVC is always represented by the worst mating conditions, that is, those obtained by subtracting the geometrical tolerance value from the minimum diameter

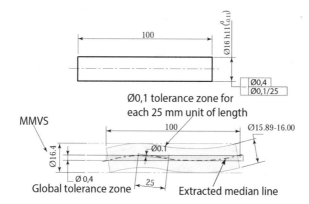

Fig. 7.12 Indications of the straightness tolerance on an axis, with specification of the total straightness and straightness per unit of length: the extracted median line should remain within a 0.4 mm diameter tolerance zone for the entire length of 100 mm and within a 0.1 mm zone for each 25 mm of length. Each circular element of the surface should fall within the prescribed dimensional tolerance

7.2.3 Straightness Tolerance in the ASME Standards

When a straightness control is applied to a cylindrical surface, by default, according to the ASME Y14.5 standard, the envelope requirement is invoked (Rule #1), and the form of the entire feature should not violate the boundary of perfect form (as in Fig. 7.5). However, it is necessary to pay particular attention, because whenever the straightness tolerance is applied to a derived median line, Rule #1 is no longer applicable, that is, the component does not have a perfect form at the maximum material. The straightness tolerance frame in Fig. 7.13 is associated with a cylindrical

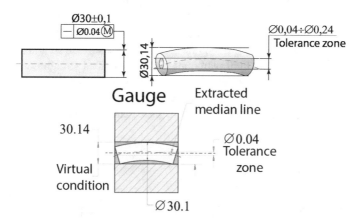

Fig. 7.13 Whenever straightness is specified on an MMC basis, functional gauging techniques may be used

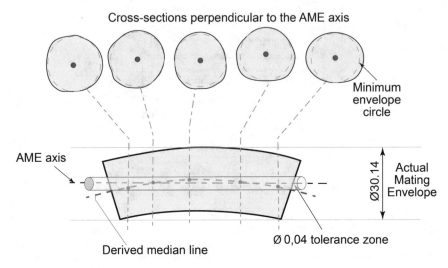

Fig. 7.14 The concept of derived median line for the case of the shaft of the previous figure. The median line is obtained from a set of central points of all the cross sections of the feature. These cross sections are perpendicular to the axis of the smallest restricted cylinder (Actual Mating Envelope)

feature of size, and the straightness control therefore applies to the derived median line.

The application of MMC is helpful since the virtual condition defines the fixed size of the functional gauge that should be used for the verification of a straightness error.

Figure 7.14 shows the concept of a derived median line in the ASME standards, obtained from the set of central points of the singular perpendicular sections of the axis of the smallest restricted cylinder (AME): the derived median line should fall within a cylinder centred on the nominal axis of an envelope of perfect form.

Figure 7.15 shows the use of a straightness tolerance on a flat surface to control line elements in multiple directions. Each line element of the surface should lie between two parallel lines separated by the amount of the prescribed straightness tolerance and in a direction indicated in the orthographic view or by the supplemental geometry of the model. In ASME Y14.5:2018, the supplementary geometry in the annotated model makes the use of the intersection plane (used in the ISO standard) unnecessary.

7.3 Flatness Tolerance

Flatness represents the condition of a surface which has all its points belonging to the same plane: the flatness error is constituted by the deviation of the real surface points from the plane. A flatness tolerance specifies a three-dimensional zone, determined by two parallel planes at a distance that is equal to the flatness control tolerance

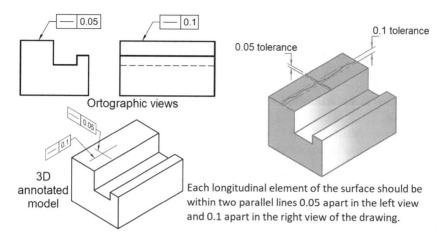

Ortographic views

3D
annotated
model

Each longitudinal element of the surface should be
within two parallel lines 0.05 apart in the left view
and 0.1 apart in the right view of the drawing.

Fig. 7.15 Straightness tolerance on a flat surface used to control line elements in multiple directions.
In the ASME Y14.5:2018 standard the supplementary geometry in the annotated model avoids the
need to use the intersection plane of the ISO standards

value. The symbol that should be used in the frame is a parallelogram, with its sides
inclined at an angle of 60° (Fig. 7.16).

Figure 7.17 shows an example of a component to which a flatness tolerance of 0.1
has been applied: in order to pass the control, the entire surface should fall between
two parallel planes 0.1 mm apart. In this case, the flatness tolerance is included in
the size tolerance of 0.2 mm.

It is not possible to apply the maximum or minimum material requirement to a
flatness tolerance, as the form tolerance controls all the points of a surface, which is
not a sizable feature.

Fig. 7.16 Flatness symbol

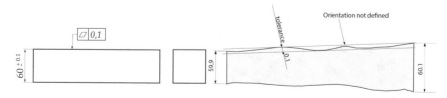

Fig. 7.17 Indication of a flatness tolerance and its interpretation

7.3.1 Flatness Parameters

The ISO 12781 standard defines the terms and concepts necessary to define specification operators (according to ISO 17450-2) for the flatness of integral features.

According to this standard, the extracted surface is a digital representation of the real surface. The flatness surface is an extracted surface intentionally modified by a filter. In the evaluation of the flatness deviation of an integral feature with a given tolerance, the flatness surface should fall between two planes, that are distant from each other by a value that is less than or equal to the specified tolerance value.

When determining the orientation of the tolerance zone, it is necessary to establish a reference plane, that is, an associated plane that fits the flatness surface according to the specified conventions, to which the deviations from flatness and the flatness parameters refer. The ISO 12781-1 technical specification considers two procedures for the determination of the reference plane:

- The **minimum zone reference planes** method (MZ), which best satisfies the tolerance zone definition (Fig. 7.18) with two parallel planes that enclose the flatness surface and have the least separation.
- The **least squares reference plane** method (LS), which provides a good approximation of the flatness deviation (Fig. 7.19), although overestimating it, and is currently the most frequently used method in coordinate measuring machines.

Local flatness deviation is defined as the deviation of a point on a flatness surface from the reference plane, in a direction normal to the reference plane.

Some of the terms related to flatness parameters are presented hereafter:

(1) **Peak-to-valley flatness deviation**, which is the value of the largest positive local flatness deviation that is added to the absolute value of the largest negative local flatness deviation. The GT modifier is used in specifications to indicate

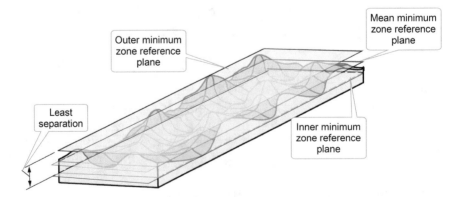

Fig. 7.18 The minimum zone reference plane method with two parallel planes that enclose the flatness surface with the least separation

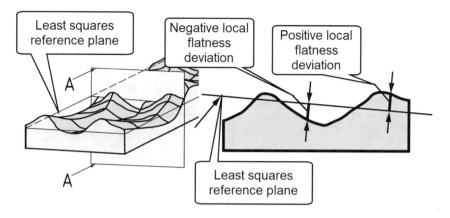

Fig. 7.19 Least squares reference plane

that a form tolerance applies to the peak-to-valley deviation relative to the least squares reference element.

(2) **Peak-to-reference flatness deviation**, which is the value of the largest positive local flatness deviation from the least squares reference line.

(3) **Reference-to-valley flatness deviation**, which is the absolute value of the largest negative local flatness deviation from the least squares reference line.

Figure 7.20 shows the application of a flatness tolerance to a flat surface (feature integral). As the principle of independency comes into force by default, the following rules are valid:

• the maximum extreme boundary MMVC is obtained by summing the maximum material dimension with the flatness tolerance.

• The flatness tolerance may be larger than the associated dimensional tolerance and each local dimension should fall within the limits of the dimensional tolerance.

Fig. 7.20 Application of a flatness tolerance to a flat surface of a prismatic component. As the principle of independency comes into force by default, the MMVC is obtained by summing the maximum material dimension with the flatness tolerance, which may be greater than the dimensional tolerance

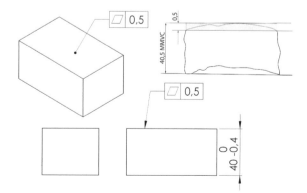

Fig. 7.21 When the
envelope requirement
(circled E symbol next to the
diameter value) is applied,
the flatness should be perfect
when the component is
produced at the maximum
material size and the
geometrical tolerance should
always be less than the
dimensional one

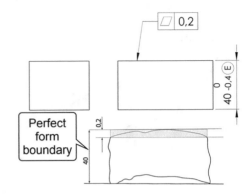

When the envelope requirement (circled E symbol next to the value of the diameter) is applied, the flatness should be perfect when the component is produced at the maximum material condition (Fig. 7.21) and the geometrical tolerance should be less than the size tolerance value. It should be recalled that this case constitutes the default condition in the ASME standards. Design experience led to the conclusion that it is advisable to prescribe a flatness tolerance that is no greater than half of the associated tolerance dimension. A flatness tolerance is usually adopted to qualify a primary datum feature.

7.3.2   Flatness Control

Several different methods may be used to check a flatness specification. For example, a flatness error can be controlled by moving a dial indicator over the entire surface, and the error is revealed as the difference between the maximum and minimum obtained measurements (Fig. 7.22a). The orientation of the plate is obtained through the use of opportune gauge blocks, and the measurement should be repeated continuously; in fact, flatness does not control a geometrical error of orientation. This method may be time consuming since the metrologist should avoid the influence of the orientation on the measurement.

Fig. 7.22 Inspecting
flatness: **a** control of the
flatness error by means of an
adjustable support or **b** by
means of a dial indicator
placed below the surface
plate

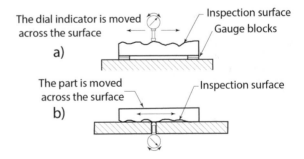

In order to avoid this problem, it is possible to use the method shown in Fig. 7.22b, that is, the part is placed on a surface plate that has a small hole. The dial indicator is placed into the small hole below the surface plate, and the part is moved in all directions. Unfortunately, if the surface is convex, it is difficult to determine the minimum indicator reading over the entire surface.

The measurement accuracy of the flatness defect can be improved by using a coordinate measuring machine (CMM) with a suitable fitting criterion. The Minimum Zone method of evaluating flatness is the most accurate and one that is best able to satisfy the ISO and ASME standards. In this case, the software of the CMM creates two theoretical parallel planes, which sandwich the extracted points as tightly as possible, and then calculates the distance between them.

7.3.3 Flatness Tolerance in the ASME Standards

In the case of the ASME standard, since the envelope requirement (Rule #1) is called into force by default, the use of the flatness tolerance has the purpose of limiting the error when the workpiece is produced with dimensions close to the least material condition (Fig. 7.23). Basically, the flatness specification further restricts the form control provisions of Rule #1.

The ASME standard of 2009 introduced the possibility of using the flatness tolerance to control the error of a derived median plane. In this case, the frame is applied to the height dimension of the workpiece, and it is possible to use the maximum material conditions (Fig. 7.24). The derived median plane should be contained between two parallel lines 0.04 apart, and each surface feature should fall within the dimensional tolerance. Figure 7.25 shows a typical application for the control of a curvature error

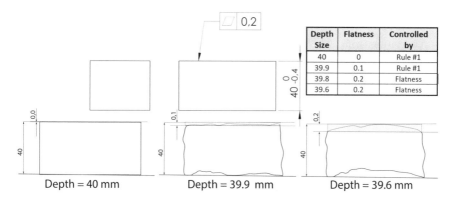

Depth Size	Flatness	Controlled by
40	0	Rule #1
39.9	0.1	Rule #1
39.8	0.2	Flatness
39.6	0.2	Flatness

Depth = 40 mm   Depth = 39.9 mm   Depth = 39.6 mm

Fig. 7.23 In the case of the ASME standard, since the envelope requirement comes into force by default, the use of the flatness tolerance has the purpose of limiting the error when the workpiece is produced with dimensions close to the least material size. The flatness tolerance zone falls within the dimensional error

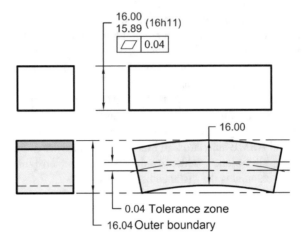

Fig. 7.24 Use of the flatness tolerance to control an error of a derived median plane. In this case, the tolerance frame is applied to the height dimension of the workpiece, and it is possible to use the MMC modifier

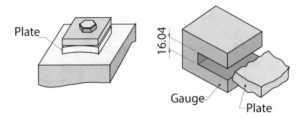

Fig. 7.25 By applying the flatness tolerance to the thickness dimension of the plate, it is possible to control the flatness error of the derived median plane. The control gauge is obtained from the virtual conditions (16 + 0.04 mm)

of a plate. By applying the flatness tolerance to the thickness dimension of the plate, it is possible to control the flatness error of the derived median plane. A functional gauge is obtained from the virtual conditions (16 + 0.04 mm).

7.4 Roundness

Roundness ('circularity' in the ASME standards) is a property of a circle. Roundness, to be more precise, is the condition at which all the points of a revolution surface are equidistant from the axis, at each section perpendicular to a common axis. A roundness error occurs when the sections made perpendicular to the axis of a round symmetry workpiece (which should nominally be circumferences) are oval, elliptic or in some way irregular.

A roundness tolerance specifies a bi-dimensional zone limited by two coaxial circles placed at a distance which is radially equal to the specified tolerance.

For cylindrical features, roundness applies for cross-sections perpendicular to the axis of the toleranced feature. For spherical features, roundness applies for cross-sections that include the centre of the sphere. A direction feature should always be indicated for revolute surfaces that are neither cylindrical nor spherical.

The example in Fig. 7.26 shows the case of indications of a roundness tolerance of 0.2 on a cylindrical workpiece; the tolerance zone falls between two concentric circles 0.2 mm apart. It should be noted that the tolerance zone is established relative to the dimensions of a transversal section, and as the ISO 8015 standard comes into force by default (and therefore the principle of independency), the dimensional and geometrical tolerances lead to an extreme boundary condition of 10.2 mm (10 + 0.2). The following rules are valid for the correct interpretation of drawings:

(1) the control of roundness is applied to all the cross sections along the toleranced feature;
(2) the roundness error may be greater than the dimensional tolerance applied to the corresponding diametric dimension;
(3) each local dimension should fall within the dimensional tolerance limits;
(4) a Maximum Material Virtual Size (MMVS) is obtained.

In the same way as for the flatness case, it is advisable, as a general rule, to choose a tolerance of a value that is less than half that of the dimensional tolerance; moreover, it is not possible to apply a maximum material or least material modifier, in that the roundness tolerance controls the boundary points of a transversal surface, a feature that cannot be sized.

The roundness error may be associated with conic surfaces or surfaces of any form, but always on condition they have round sections, as shown in Fig. 7.27.

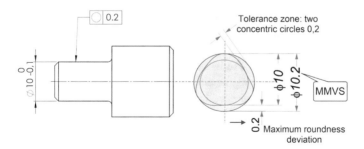

Fig. 7.26 Indication and interpretation of a roundness tolerance

Fig. 7.27 Roundness
tolerance applied to an
axial-symmetric workpiece.
The extracted
circumferential line should
be contained between two
circles (0.04 mm apart) in
any cross-section
perpendicular to datum axis
A, as indicated by the
direction feature symbol

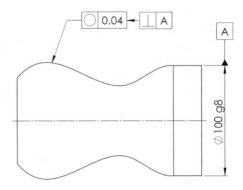

7.4.1 Roundness Parameters

The ISO 12181 standard defines the terms and concepts necessary to define
specification operators (according to ISO 17450-2) for the roundness of integral
features.

According to this standard, the extracted circumferential line is a digital representation of the intersection of the real surface and a roundness plane while the
roundness profile is an extracted circumferential line intentionally modified by a
filter. The roundness plane is a plane that is perpendicular to the roundness axis
for the full extent of the feature, while the roundness axis is the axis of a feature
associated with an integral feature.

When evaluating the roundness deviation of an integral feature with a given tolerance, the roundness profile should fall between two circles that are distant from each
other by a value that is less than or equal to the specified tolerance value.

When determining the roundness deviation, it is necessary to establish a reference
circle, that is, an associated circle that fits the roundness profile according to specified
conventions, to which the deviations from roundness and the roundness parameters
refer. The inspection of roundness, by means of CMM measuring machines, foresees,
starting from the extracted line, the association of a reference circle, with respect to
which two concentric circles can be defined in order to calculate the magnitude of
the tolerance zone that contains the circle itself.

The ISO 12181-1 technical specification considers four procedures that can be
used to determine the reference circle:

- The **minimum zone reference circles** method (MZ), whereby two concentric circles enclose the roundness profile and have the least radial separation
  (Fig. 7.28).
- The **least squares reference circle** (LS), that is, a circle in which the sum of the
  squares of the local roundness deviations is a minimum (Fig. 7.29).
- The **minimum circumscribed reference circle** (MC), that is, the smallest
  possible circle that can be fitted around the roundness profile (Fig. 7.30).

Fig. 7.28 The minimum zone reference circle method with two concentric circles which enclose the roundness profile and have the least radial separation. The roundness error is the radial distance between the two circles

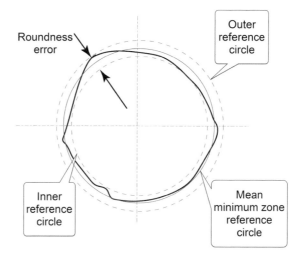

Fig. 7.29 Least square reference circle. The roundness error is expressed as the sum of the largest positive local roundness deviation and the absolute value of the largest negative local roundness deviation

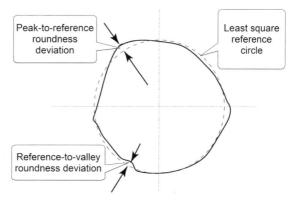

Fig. 7.30 The minimum circumscribed reference circle (MC), that is, the smallest possible circle that can be fitted around the roundness profile. The roundness error is the largest local deviation from the reference circle

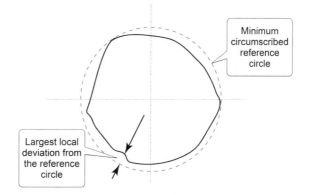

Fig. 7.31 The maximum
inscribed reference circle
(MI), that is, the largest
possible circle that can be
fitted within the roundness
profile. The roundness error
is the largest local deviation
from the reference circle

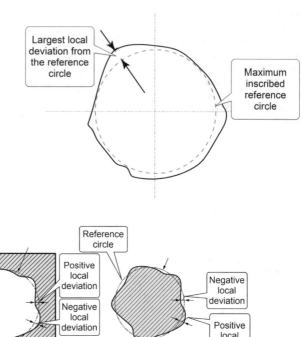

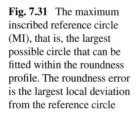

Fig. 7.32 The local form deviation of an internal feature and of an external roundness feature

- The **maximum inscribed reference circle** (MI), that is, the largest possible circle that can be fitted within the roundness profile (Fig. 7.31).

  Local roundness deviation is defined as the minimum distance between a point on a roundness profile and the reference circle (Fig. 7.32).

  Some of the terms related to roundness parameters are presented hereafter:

(1) **Peak-to-valley roundness deviation**, which is the value of the largest positive local roundness deviation added to the absolute value of the largest negative local roundness deviation. The GT modifier is used in specifications to indicate that a form tolerance applies to the peak-to-valley deviation relative to the least squares reference element.

(2) **Peak-to-reference roundness deviation**, that is, the value of the largest positive local roundness deviation from the least squares reference line.

(3) **Reference-to-valley roundness deviation**, that is, the absolute value of the largest negative local roundness deviation from the least squares reference line.

The control of roundness may be achieved by means of coordinate measuring machines, or through the use of other systems, such as the roundness checking machine shown in Fig. 7.33, which is composed of three main features, a high-precision spindle, a probe and a computer.

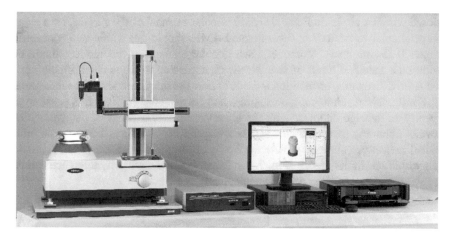

Fig. 7.33 Measurements may be carried out by means of a roundness checking machine, such as the one shown here (Mitutoyo RA-1600 M Roundness Tester)

Fig. 7.34 Specification using the least squares (Gaussian) reference feature specification element and the valley depth characteristic specification element

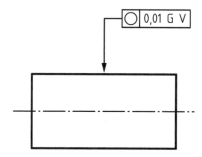

In the past, the definition of the reference circle was left to the discretion of the measurement technicians who, on many occasions, applied the GPS default, that is, the minimum zone criterion (Chebyshev). However, with the new ISO 1101 standard of 2017, it is possible to apply a further series of criteria that can be adopted for all form controls and which can be identified by means of modifiers, as shown in Fig. 7.34.

7.4.2 Roundness Tolerance in the ASME Standards

In the case of the ASME standards, since the envelope requirement is valid by default, the use of circularity tolerance has the purpose of limiting the error when the workpiece is produced with dimensions close to the least material condition (see Fig. 4.25). The geometrical tolerance should always be less than the dimensional tolerance of the inspected feature, and the same effect is obtained with ISO using the symbol E closed within a circle. In Fig. 7.35, each circular element of the surface, on a plane perpendicular to the axis of the unrelated AME, should fall within two concentric circles 0.25 mm apart. Moreover, each circular element of the surface should be within the specified limits of size. However, according to the general principles of the ASME standards, no mention is made of how the roundness should be measured.

For flexible revolute components in a non-restrained condition, it is possible to specify an average diameter with the abbreviation "AVG", that is, the average of several diametric measurements across a circular or cylindrical feature. The individual measurements may violate the limits of size, but the average value should fall within the limits of size (Fig. 7.36). Moreover, the circularity tolerance can be greater than the size tolerance on the diameter.

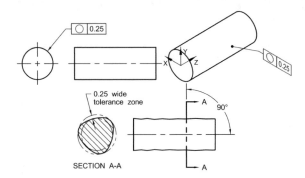

Fig. 7.35 Each circular element of the surface, on a plane perpendicular to the axis of the unrelated AME, should fall within two concentric circles 0.25 mm apart

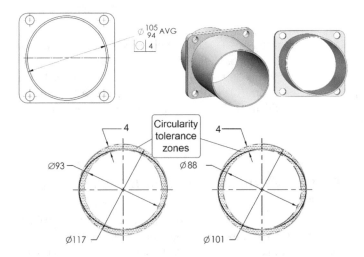

Fig. 7.36 Specifying Circularity with the average diameter

7.5 Cylindricity

Although various sections with planes perpendicular to the axis of a workpiece are circumferences, there may be differences in diameter among them; cylindricity is in fact a condition of a revolution surface in which all the points of the surface are equidistant from a common axis.

The symbol that should be placed in the tolerance frame to indicate a cylindrical tolerance is shown in Fig. 7.37.

A real cylindrical surface can be subject to deviation errors from the cylindrical form as a combination of simple elements, caused by machining errors and/or distortions resulting from thermal, pressure or stress effects, tool wear, and/or vibrations. These deviations may be classified as (Fig. 7.38):

- **Median line deviation**: the deviation of a nominal cylindrical workpiece which has a curved axis, but a circular and constant radius cross-section.
- **Radial deviations**, i.e. variations in the cross-section dimension: the deviation of a nominal cylindrical workpiece which has all the cross-sections circular and concentric to a straight axis, but whose diameters vary along the axis according to simple or complex laws or even randomly (typical deviations include conical, barrel or more complex forms).
- **Cross-section deviations**: the deviation on a nominal cylindrical workpiece that has cross-sections of the same size and form, but which are not round.

A cylindrical tolerance specifies a three-dimensional zone between two coaxial cylinders within which the surfaces should fall. The example in Fig. 7.39 shows a

Fig. 7.37 The cylindricity symbol

Fig. 7.38 Deviations from cylindrical form

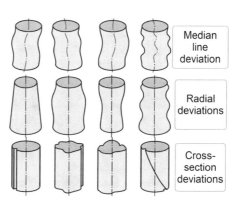

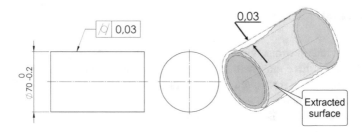

Fig. 7.39 Cylindrical tolerance and its interpretation

cylindrical shaft with a cylindricity tolerance of 0.03 mm, which is measured in a radial direction, that is, through coaxial cylinders 0.03 apart.

Since a cylindricity tolerance can be interpreted as a roundness tolerance that is extended over the entire cylindrical surface, it can be used to control the roundness, the straightness and the taper of the surface features simultaneously.

Again in this case, it is advisable to choose a form tolerance of a value that is less than half of the dimensional tolerance; moreover, the maximum material requirement cannot be applied, as a cylindrical surface is not a sizeable feature. Cylindricity can be inspected in a similar manner to the way roundness is inspected, that is, utilising a coordinate measuring machine or a specific measuring machine, as shown in Fig. 7.33.

Together with the control of roundness, the example in Fig. 7.40 also shows an example of an indication of cylindricity tolerance of 0.1 on a cylindrical work-piece; the tolerance zone falls between two coaxial cylinders 0.1 mm apart. It should be noted that, since the independency principle is called into force by default, the dimensional and geometrical tolerances have a virtual condition of 10.1 mm (10 + 0.1).

The following rules are valid for a correct interpretation of drawings:

(1)   the cylindricity error may be greater than the dimensional tolerance applied to the corresponding diametric dimension;

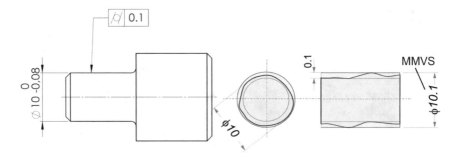

Fig. 7.40 Interpretation of the independency principle for a cylindricity tolerance. A maximum material virtual size (MMVS) of 10.1 mm is obtained

(2) each local dimension should fall between the dimensional tolerance limits;
(3) an extreme boundary of the maximum material (MMVS) is obtained.

7.5.1 Cylindricity Parameters

The ISO 12180 standard defines the terms and concepts necessary to define specification operators (according to ISO 17450-2) for the cylindricity of integral features.

According to this standard, the extracted surface is a digital representation of the real surface, while the cylindricity surface is an extracted surface (of a cylindrical type) intentionally modified by a filter.

The generatrix plane is a half plane that passes through the axis of the associated cylinder, while the extracted generatrix line is a digital representation of the intersection line of the real surface and a generatrix plane.

When evaluating the cylindricity deviation of an integral feature with a given tolerance, the cylindricity surface should fall between two cylinders that are distant from each other by a value that is less than or equal to the specified tolerance value.

When determining the cylindricity deviation, it is necessary to establish a reference cylinder, that is an associated cylinder that fits the cylindricity surface according to specified conventions, to which the deviations from cylindrical form and the cylindricity parameters refer. The associated derived axis of a cylindrical feature is the axis of the reference cylinder(s).

The inspection of cylindricity, by means of CMM measuring machines, foresees, starting from the extracted surface, the association of a reference cylinder, with respect to which two coaxial cylinders can be defined in order to calculate the magnitude of the tolerance zone.

The ISO 12180-1 technical specification considers four procedures that can be used to determine the reference cylinder:

- The minimum zone reference cylinders method (MZ), that is, two coaxial cylinders enclose the cylindricity surface[2] and have the least radial separation.
- The least squares reference cylinder (LS), that is, a cylinder for which the sum of the squares of the local cylindricity deviations is the minimum value.
- The minimum circumscribed reference cylinder (MC), that is, the smallest possible cylinder that can be fitted around the cylindricity surface.
- The maximum inscribed reference cylinder (MI), that is, the largest possible cylinder that can be fitted within the cylindricity surface.

Local cylindricity deviation is defined as the deviation of a point on a cylindricity surface from the reference cylinder, the deviation being normal to the reference cylinder (Fig. 7.41). The local generatrix deviation is a deviation of a point on a

---

[2]According to Sect. 3.2.2 of ISO 12180–1 standard, a cylindricity surface is an extracted surface (type cylinder) intentionally modified by a filter.

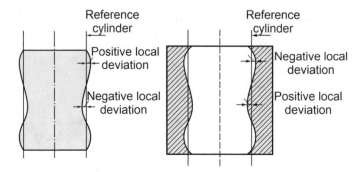

Fig. 7.41 Local form deviations of an external and internal cylindrical feature

generatrix from the reference line, the deviation being normal to the reference line.

The local generatrix straightness deviation is a straightness deviation value calculated from a generatrix profile obtained from an intersection of a plane through the axis of the least squares reference cylinder and the extracted cylindrical feature (Fig. 7.42).

The absolute value $|D_1 - D_2/|2$ is the local cylinder taper value. This parameter is usually evaluated with a length L of 100 mm.

Some of the terms related to cylindricity parameters are given hereafter:

(1) **Peak-to-valley cylindricity deviation**, which is the value of the largest positive local cylindricity deviation added to the absolute value of the largest negative local cylindricity deviation. The GT modifier is used in specifications to indicate that a form tolerance applies to the peak-to-valley deviation relative to the least squares reference element.

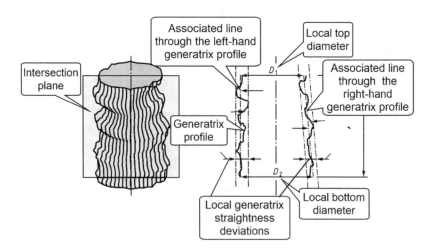

Fig. 7.42 Local generatrix straightness deviations

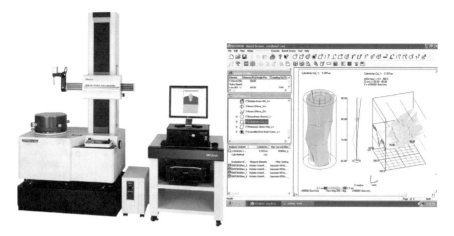

Fig. 7.43 Cylindricity inspection with a CNC form measuring instrument (Roundtest Extreme RA-H5200CNC, Mitutoyo)

(2) **Peak-to-reference cylindricity deviation**, which is the value of the largest positive local cylindricity deviation from the least squares reference cylinder.

(3) **Reference-to-valley cylindricity deviation**, which is the absolute value of the largest negative local cylindricity deviation from the least squares reference cylinder.

The control of cylindricity may be achieved by means of coordinate measuring machines, or through the use of other systems, such as the "Roundtest Machine" shown in Fig. 7.43, which is a CNC form measuring instrument that combines high accuracy with automatic CNC measurements.

Again in this case, it is possible to indicate the reference cylinder association methodology, from which it is possible to derive two coaxial cylinders, which can be used to obtain the cylindricity measurement. The association shown in Fig. 7.44 is in fact obtained with the smallest circumscribed cylinder (N and X modifiers).

7.5.2 Cylindricity Tolerance in the ASME Standards

In the case of the ASME standard, since the envelope requirement (Rule #1) is called into force by default, the use of the cylindricity tolerance has the purpose of limiting the error when the workpiece is produced with dimensions close to the least material condition (Fig. 7.45). Basically, the cylindricity specification further restricts the form control provisions introduced by Rule #1.

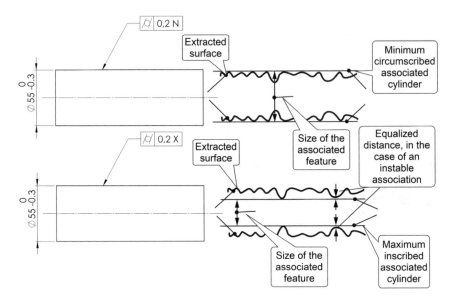

Fig. 7.44 Specification of the reference cylinder association methodology

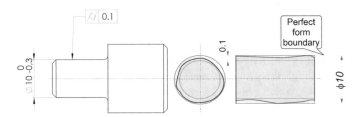

Fig. 7.45 Indication and interpretation of a cylindricity error in the ASME standards. The geometrical tolerance should always be less than the dimensional tolerance of the inspected feature, and the same effect can be obtained with ISO using the symbol E enclosed within a circle

Reference

1. Krulikowski A (2010) Alex Krulikowski's ISO geometrical tolerancing guide. Effective Training Inc.

Chapter 8
Orientation Tolerances

Abstract Orientation tolerances (parallelism, perpendicularity and angularity) are used to control the orientation of a feature (surface or feature of size) with respect to one or more datums. When the orientation tolerances are applied to a feature of size and an MMR or LMR is added, the control of the orientation deviation no longer refers to the median line, but to the entire extracted feature (MMVC boundary), and it should not violate the MMVC virtual condition. The ISO and ASME standards use two different approaches to control the orientation of a feature of size: in order to orient a feature of size, the ISO standards define the concept of extracted median line or median surface. Instead, in the ASME standards, the axis or median plane is used to control the orientation of a feature of size.

8.1 Introduction

Orientation tolerances (parallelism, perpendicularity, angularity) control the orientation of a feature (surface or feature of size) with respect to one or more datums.

A parallelism control defines the deviation of a part feature from parallelism, and it is used for geometries that are at 180° with respect to another geometry.

A perpendicularity control defines the deviation of a part feature from perpendicularity, and it is used for geometries at 90° from each other.

An angularity control defines the deviation of a part feature with respect to a determined inclination, and it is used for geometries that are neither at 180° nor at 90° from each other.

8.2 Parallelism

A parallelism control defines the deviation of a part feature from parallelism, and it is used for geometries that are at 180° from each other. This type of error can be applied to a derived median line or a surface, and its symbol is two parallel dashes inclined at 60°.

The toleranced feature can be an integral feature or a derived feature. The theoretically exact dimension (TED) angles that are locked between the nominal toleranced feature and the datums should be defined by means of implicit TEDs (0°).

8.2.1 Parallelism of a Median Line Related to a Datum System

8.2.1.1   Parallelism of a Median Line Related to a Datum Axis

Since the tolerance value in the parallelism control in Fig. 8.1 is not preceded by the Ø symbol, and being an indicator of an orientation plane, the median line extracted from the upper hole should fall between two planes 0.1 mm apart, parallel to datum A, and with the orientation specified by datum B (Fig. 8.2).

In previous practice, as an alternative, the orientation of the tolerance zones was defined with a secondary datum (Fig. 8.3). However, a more rigorous parallelism control is obtained, as indicated in Fig. 8.4, when the extracted median falls between *two pairs of parallel planes*, which are parallel to datum axis A, and positioned 0.1 and 0.2 apart, respectively. The orientation of the planes that limit the tolerance zones is specified, with respect to datum plane B by the orientation plane indicators.

The tolerance zone in Fig. 8.5 is limited by a 0.3 mm diameter cylinder, whose axis is parallel to datum A, because the tolerance value is preceded by the Ø symbol.

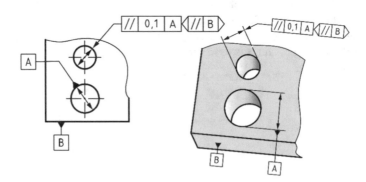

Fig. 8.1 Parallelism specification of a median line related to a datum axis; the planes that limit the tolerance zone are parallel to datum plane B, as specified by the orientation plane indicator

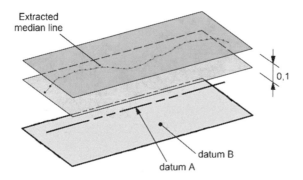

Fig. 8.2 Control of the extracted median line shown in the previous figure; in this case, the extracted median line of the upper hole should fall within two planes 0.1 mm apart, parallel to datum A and with the orientation specified by datum B, as specified by the orientation plane indicator

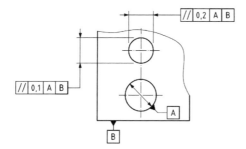

Fig. 8.3 A previous practice, alternative to the indication of the orientation plane shown in Fig. 8.1

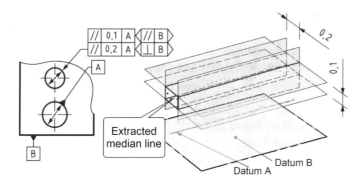

Fig. 8.4 Parallelism indication with two orientation plane indicators

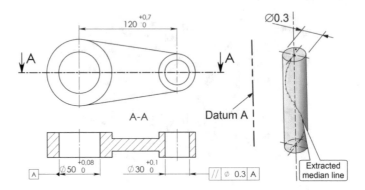

Fig. 8.5 The use of the Ø symbol defines a cylindrical tolerance zone parallel to the datum within which the extracted medium line should fall

Fig. 8.6 If the tolerance value is not preceded by the Ø symbol, the extracted median line should fall between two parallel planes 0.01 mm apart, which are parallel to datum plane B

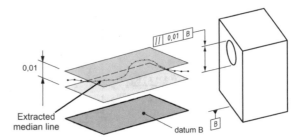

8.2.1.2 Parallelism of a Median Line Related to a Datum Plane

A parallelism specification of a median line related to a datum plane, as shown in Fig. 8.6, where the extracted median line should fall between two parallel planes 0.01 mm apart, which are parallel to datum plane B.

8.2.2 Parallelism of a Set of Lines on a Surface Related to a Datum Plane

Each extracted line in Fig. 8.7, parallel to datum plane B, as specified by the *intersection plane indicator*, should fall between two parallel lines 0.2 apart, which are parallel to datum plane A. The tolerance zone is limited by two parallel lines 0.2 mm apart and oriented parallel to datum plane A, that is, the lines lying on a plane parallel to datum plane B.

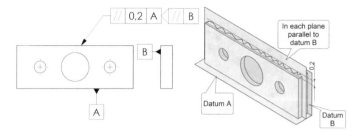

Fig. 8.7 Parallelism specification of a set of lines on a surface

8.2.3 Parallelism of a Planar Surface Related to a Datum System

Figure 8.8 shows a parallelism specification of a planar surface related to a datum plane; the upper extracted surface should fall between two parallel planes 0.4 mm apart, which are parallel to datum plane A. It should be noted that the parallelism control also limits the flatness of the surfaces.

In this case, the principle of independency is also invoked by default for orientation tolerances, and the following rules are therefore valid (Fig. 8.9):

(1) the parallelism tolerance can have a higher value than the size tolerance value;
(2) each local dimension measured between two points should fall within the dimensional limits;
(3) the form deviations should fall within the parallelism tolerance values.

8.3 Perpendicularity

The toleranced feature controlled by a perpendicularity specification may be either an integral feature or a derived feature. The theoretically exact dimension (TED) angles that are locked between the nominal toleranced feature and the datums should be

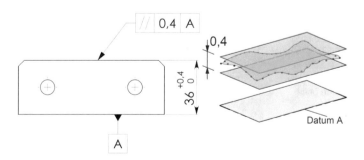

Fig. 8.8 Parallelism specification of a planar surface related to a datum plane

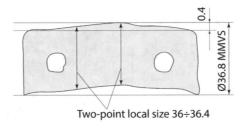

Two-point local size 36÷36.4

Fig. 8.9 Interpretation of the parallelism specification of the previous figure. The orientation toler-ance is applied to a feature of size, and the principle of independency is thus invoked by default. Therefore, the parallelism tolerance may have a higher value than the size tolerance, and the form deviations (flatness in this case) should fall within the indicated tolerance values [1]

defined by means of implicit TEDs (90°). The control of perpendicularity (symbol-ised by two orthogonal dashes) is generally used to qualify a secondary or tertiary datum feature.

8.3.1 Perpendicularity of a Median Line Related to a Datum System

In the case shown in Fig. 8.10, the extracted median line of the cylinder should fall between two pairs of parallel planes, perpendicular to datum plane A, and posi-tioned 0.1 mm and 0.2 mm apart, respectively. The orientation of the planes that limit the tolerance zones is specified, with respect to datum plane B, by the orientation plane indicators. In short, the orientation of the planes is defined by two orienta-tion plane symbols with respect to datum B. One couple of planes (0.1 mm apart) is perpendicular to datum B and the other (0.2 mm apart) is parallel to datum B (Fig. 8.11).

Fig. 8.10 Perpendicularity specification of a median line related to a datum system; the extracted median line of the cylinder should fall between two pairs of parallel planes, perpendicular to datum plane A, as indicated by the orientation plane indicators

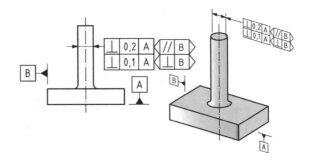

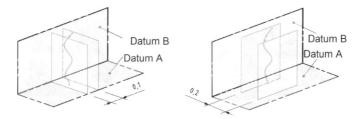

Fig. 8.11 The tolerance zone is limited by two pairs of parallel planes 0.1 mm and 0.2 mm apart, respectively, and perpendicular to each other. Both planes are perpendicular to datum A. One pair of planes is perpendicular to datum B and the other is parallel to datum B

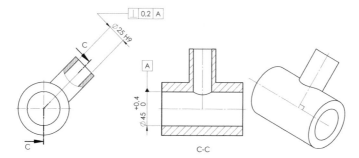

Fig. 8.12 Perpendicularity specification of a median line related to a datum straight line; the extracted median line of the 25 mm diameter hole is controlled by a perpendicularity tolerance with respect to the axis of the horizontal hole, which is taken as a datum (it should be noted that, in the lateral view, the two axes are not on the same vertical plane)

8.3.2   Perpendicularity of a Median Line Related to a Datum Straight Line

As shown in Fig. 8.12, the extracted median line of the 25 mm diameter hole is controlled by a perpendicularity tolerance with respect to the axis of the horizontal hole, which is taken as a datum. Since the tolerance value is not preceded by the Ø symbol, the extracted median line should fall between two parallel planes 0.2 mm apart, which are perpendicular to datum axis A (Fig. 8.13).

8.3.3   Perpendicularity of a Median Line Related to a Datum Plane

The tolerance zone defined by the specification in Fig. 8.14 is limited by a 0.1 mm diameter cylinder, whose axis is perpendicular to the datum A, because the tolerance value is preceded by the Ø symbol.

Fig. 8.13 Since the
tolerance zone is not
preceded by the Ø symbol,
the tolerance zone is limited
by two parallel planes
0.2 mm apart and
perpendicular to datum axis
A

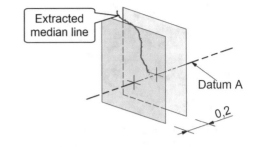

Fig. 8.14 Perpendicularity
tolerance applied to a
cylindrical feature of size
utilising the Ø symbol

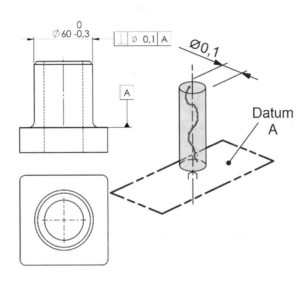

8.3.4 Perpendicularity of a Planar Surface Related to a Datum Plane

The extracted surface in Fig. 8.15 should be contained between two parallel planes
0.1 mm apart, which are perpendicular to datum plane A. Again, in this case, flatness
is implicitly controlled. The rotation of the tolerance zone around the normal of

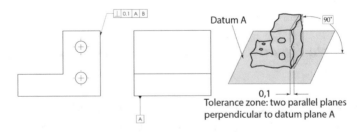

Fig. 8.15 Perpendicularity specification of a planar surface related to a datum plane

the datum plane is not defined with the indication given in Fig. 8.15, and only the direction is specified. The assembly and control procedures of the workpiece are also illustrated in Fig. 8.16: the component is free to rotate around a horizontal axis, and this instability could influence the verification of the perpendicularity deviation. In order to avoid this problem, the perpendicularity tolerance *may be prescribed with two datums.*

As a result, in the case of Fig. 8.17, the perpendicularity tolerance is applied with two datum planes, because the assembly of the workpiece implies the alignment with surface B; the control takes place by first placing the workpiece on the primary plane and then on the secondary plane (Fig. 8.18).

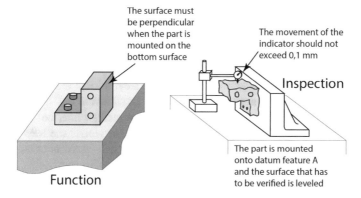

Fig. 8.16 Functional requirements and inspection of a workpiece subject to a perpendicularity requirement

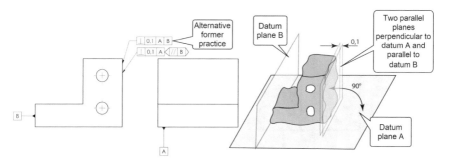

Fig. 8.17 Drawing of a workpiece subject to a perpendicularity requirement with two datum planes. The extracted surface should fall between two parallel planes 0.1 mm apart, which are perpendicular to datum plane A and parallel to datum B, as specified by the orientation plane indicator

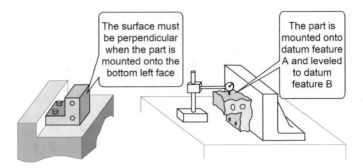

Fig. 8.18 Functional requirements and inspection of a workpiece subject to a perpendicularity requirement related to two datum features

8.3.5 *Perpendicularity (MMR) Applied to a Feature of Size*

When the perpendicularity tolerance is applied to a cylindrical feature of size and a maximum material requirement (MMR) is added, as shown in Fig. 8.19, the control of the orientation deviation no longer refers to the median line, but to the entire extracted feature (MMVC boundary), and it should not violate the MMVC virtual condition of 60.1 mm, which is calculated by summing the maximum material size (60 mm) with the perpendicularity geometrical tolerance (0.1 mm).

The MMVC represents the most unfavourable mating condition by which a designer is able to ensure full functionality and interchangeability of the produced parts, and all at a minimum cost.

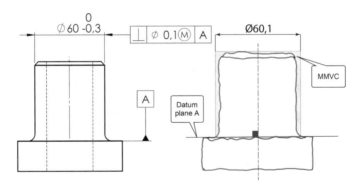

Fig. 8.19 Perpendicularity tolerance (MMR) applied to a feature of size: the control of the orientation deviation no longer refers to the median line, but to the entire extracted feature, which should not violate the MMVC virtual conditions of 60.1 mm

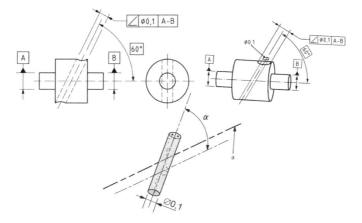

Fig. 8.20 Angularity specification of a median line related to a datum straight line; the extracted median line should be contained within a 0.1 diameter cylinder, which is inclined at a theoretically exact angle of 60° with respect to the common datum straight line A-B

8.4 Angularity

Angularity represents the conditions of a surface, or derived median line, that is at a determined angle (other than 90° and 0°) with respect to a datum. The theoretically exact dimension angles that are locked between the nominal toleranced feature and the datums should be defined by means of at least one explicit TED. Additional angles may be defined by means of implicit TEDs (0° or 90°).

8.4.1 Angularity of a Median Line Related to a Datum Straight Line

In the case of Fig. 8.20, the extracted median line should fall within a 0.1 diameter cylinder, which is inclined at a theoretically exact angle of 60° with respect to the common datum straight line A-B. The considered line and the datum line are not on the same plane.

8.4.2 Angularity of a Median Line Related to a Datum System

This is the case shown in Fig. 8.21, where the extracted median line should be within a 0.1 mm diameter cylindrical tolerance zone that is parallel to datum plane B and inclined at a theoretically exact angle of 60° with respect to datum plane A. If the tolerance value is preceded by the Ø symbol, the tolerance zone is a cylinder with the

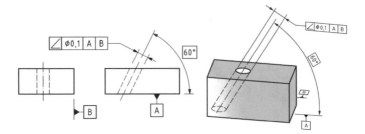

Fig. 8.21 Angularity specification for an extracted median line related to a datum system. The presence of the theoretically exact dimension (TED) can be noted for the angle

Fig. 8.22 Interpretation of the tolerance zone of the previous figure. The extracted median line should fall within a cylindrical tolerance zone (of 0.1 mm diameter), parallel to datum B and inclined by 60° with respect to datum plane A

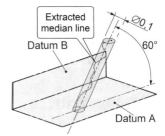

indicated diameter, and with the axis inclined with respect to the datum feature and parallel to secondary datum B (Fig. 8.22). An alternative to the specification with two datums is the orientation plane indicator.

8.4.3   Angularity of a Planar Surface Related to a Datum Plane

The tolerance zone in Fig. 8.23 is composed of two planes 0.1 mm apart and inclined by 30° with respect to datum plane A. The entire surface should fall within the tolerance, and this means that there is also a flatness control of 0.1 mm. The control

Fig. 8.23 Angularity specification of a planar surface related to a datum plane. The tolerance zone is limited by two parallel planes 0.1 mm apart and inclined, at the specified angle, to datum A

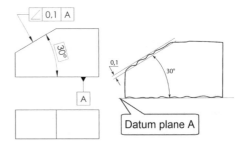

is achieved by mounting the part onto the gauge equipment, using a sine plate set at the basic angle to make the toleranced surface parallel to the surface plate. A dial indicator is used to verify that the surface elements are within the angularity tolerance zone (Fig. 8.24). Angularity does not control the position of the tolerance zone, for which it may translate in any direction, but the zone must remain oriented to datum plane A at an angle of 30°. However, the rotation of the tolerance zone around the normal to the datum plane is not defined in the drawing, thus making the inspection difficult and expensive.

In all the illustrated examples, the angular dimensions should be framed, as they represent the theoretically exact dimensions that are not subject to general tolerances and which define the theoretical inclination of the tolerance zone with respect to the datum.

Figure 8.25 shows the same workpiece as Fig. 8.23, with indication of two datum planes; the only difference from the previous case pertains to the control modality, in that, in order to avoid instability problems, the workpiece is first orientated with respect to primary plane A and then aligned according to datum plane B.

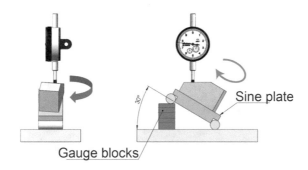

Fig. 8.24 Control with a piece of equipment called a sine plate set. It should be noted that rotation around an axis perpendicular to the datum is possible, and this leads to difficulty in the control

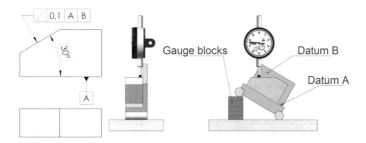

Fig. 8.25 Angularity specification of a planar surface with respect to a datum system

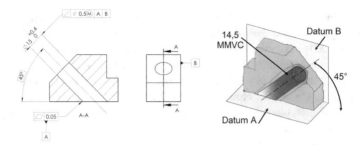

Fig. 8.26 Effect of the maximum material requirement applied to an angularity tolerance. In this case, the extracted median line is not controlled, but the edge of the hole is, and this edge should not violate the MMVS of 14.5 mm, whose axis is inclined by 45° with respect to datum A

8.4.4 Angularity (MMR) Applied to a Feature of Size

Figure 8.26 shows the effect of a maximum material requirement (MMR) applied to an angularity specification. In this case, the extracted median line is not controlled, but the MMVC (MMVS = 14.5 mm) of the hole is. The hole boundaries should not violate the MMVC, whose axis is inclined by 45° with respect to datum A. Datum B plays the role of *stabilising* the angularity verification process [1].

8.5 Orientation Tolerances in the ASME Standards

As mentioned in the previous sections, the ISO and ASME standards use two different approaches to determine or simulate an axis, above all for orientation tolerances: in fact, as has already been seen, in order to orient a cylindrical feature, the ISO standards define the concept of derived median line (or extracted median line). Instead, in the ASME standards, the axis of the unrelated AME (smallest circumscribed cylinder) is used to verify the perpendicularity control (Fig. 8.27).

In the ASME standards it is emphasised that orientation tolerances does not control the location of features, and when specifying an orientation tolerance, consideration should therefore be given to the control of orientation, as already established through other tolerances, such as position, run-out, and profile controls (Fig. 8.28).

ISO 1101 has recently introduced the new **tangent plane** symbol, which was previously present in ASME Y14.5 M of 1994. When a tangent plane symbol is specified with a geometric tolerance, a plane in contact with the high points of the feature should fall within the tolerance zone established by the geometric tolerance. In this case, the form of the toleranced feature is not controlled by the geometric tolerance and some points of the toleranced feature may lie outside the tolerance zone (Fig. 8.29).

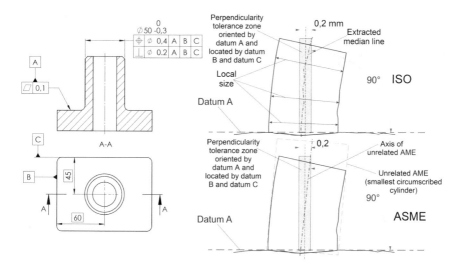

Fig. 8.27 The ISO standards define the concept of extracted median line in order to control the orientation. Instead, the ASME standard controls the orientation of the axis of the smallest circumscribed cylinder and, at the same time, the cylindrical feature should have the perfect form at the maximum material condition

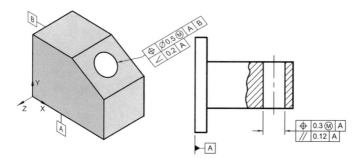

Fig. 8.28 Orientation tolerances does not control the location of features, and it is therefore necessary to specify other location controls

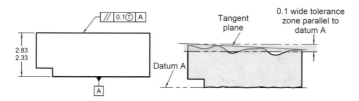

Fig. 8.29 Control of the parallelism of a planar surface with a tangent plane modifier. In this case, the form of the toleranced feature is not controlled by the geometric tolerance, and some points of the toleranced feature may lie outside the tolerance zone

Reference

1. Alex Krulikowski (2015) ISO GPS Ultimate Pocket Guide, Effective Training Inc.

Chapter 9
Location Tolerances

Abstract This chapter describes how to use location tolerances to specify an allowable location error. The location tolerances that should be used, in function of the feature of the workpiece that has to be located, are in particular presented. The effects of the material condition requirement and the correct choice of the modifiers for position tolerances (axis and surface interpretation) are illustrated. Particular emphasis is given to the location of the patterns, as introduced with the new rules of the ISO 5458:2018 standard (multiple indicator pattern specification and simultaneous requirement). When geometrical position tolerances are applied, the value of the tolerance is calculated from the mating conditions (fixed and floating fastener conditions). The ASME standards specify, without any shadow of doubt, that the position tolerance symbol should only be utilised for a "feature of size", while the ISO standards allow it to be used to position a planar surface. The ISO standards are defined as "CMM Friendly", that is, the preferred control system is the coordinate measurement machine, while the ASME standards are based on functional gauges that represent a physical representation of the tolerance zone. A special functional application of location tolerances is to control concentricity and symmetry. These controls have been removed from the new ASME standard.

9.1 Introduction

Location tolerances establish the variations that are permitted, with respect to a fixed theoretical location, of a feature of a component with respect to one or more datums. Table 9.1 indicates the location tolerances that should be used in function of the features of the workpiece that has to be located. A location tolerance may be used to locate an axis or a median plane extracted from a feature of size and, if it is necessary to locate the border of a feature of size (for example, the boundary of a hole to ensure the passage of a screw), it is possible to add a maximum or least material modifier. In order to locate a cylinder, with respect to a datum axis, apart from the location, it is possible to use both a run-out control and a profile. Finally, in order to locate a surface with respect to one or more datums, it is possible to use the control of the profile, which is dealt with in detail in the next sections.

Table 9.1 Geometrical tolerances that can be used in function of the geometrical feature of the workpiece that has to be located	Feature that has to be localised	Geometric control
	Extracted median line or surface	$\oplus$ =
	Feature of size boundary (MMVC, LMVC)	$\oplus$ with modifiers Ⓜ, Ⓛ
	Coaxial cylinder	$\oplus$ ⌓ ⫽ ⌱ ◎
	Any surface	$\oplus$ ⌓

Figure 9.1 shows an example of the positioning of some features of a component. The 6 mm diameter hole has the role of positioning, with the smallest possible error, the axis of a reference pin; as there are no modifiers, the error zone is represented by a 0.1 mm diameter cylinder, whose centre is defined by a theoretically exact dimension.

The pattern of 9 mm holes of a bolted joint (with a large diameter in order to accommodate the bolt size), is located by means of a position tolerance with a maximum material requirement, which has the objective of positioning six MMVC boundaries, with an MMVS of 8.44 mm.

The groove and the 100 mm cylindrical surface are located with a profile tolerance which indirectly controls the shape and size. Finally, the coaxiality of the 24 mm diameter hole is controlled by a run-out tolerance.

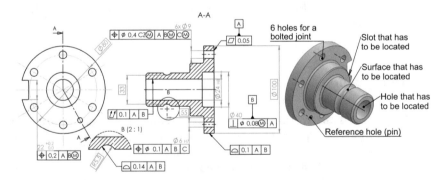

Fig. 9.1 How to control the location of some features of a component

9.2 Position Tolerances

Position tolerances are mainly used *to locate a features of size*, that is, to define a zone within which a centre point, a derived axis, a median plane or an MMVC is permitted to move from a theoretically exact position. Therefore, location tolerances make use of theoretically exact dimensions (TED) to establish the theoretical location of the features that have to be located with respect to the datum planes, and from which to control the location and orientation of the tolerance zones.

In ISO standards, the position tolerance can be applied *to planar surfaces*, cylindrical features (such as shafts and holes) and to non-cylindrical features, such as slots, ruts, ribs, etc., and the symbol that should be used is indicated in Fig. 9.2.

In order to understand the advantages of position tolerances, reference can be made to Fig. 9.3, where two different methods are shown for the positioning of a hole:

(a) coordinate tolerances, where the hole is located according to the dimensional tolerances.
(b) Position tolerance, with the use of theoretically exact dimensions.

In the first case (Fig. 9.4a), the cross section of the 3D tolerance zone may be represented by a 2D square zone, if the tolerances are the same along both the

Fig. 9.2 The position tolerance symbol

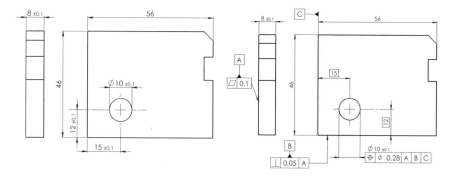

Fig. 9.3 Comparison of coordinate tolerancing and position tolerances with the use of theoretically exact dimensions

Fig. 9.4 Coordinate tolerancing: square tolerance zone (**a**), rectangular tolerance zone (**b**)

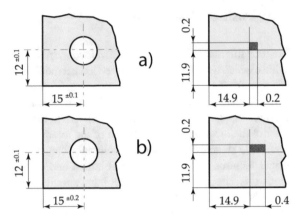

vertical and horizontal directions, or by a 2D rectangle zone, under the hypothesis of different tolerances (Fig. 9.4b). The position of a point of the extracted median line may vary inside the tolerance zone, and the maximum deviation may therefore occur along the diagonal; if the side of the square is, for example, 0.2 mm, the maximum deviation will be equal to:

$$0.2 \times \sqrt{2} = 0.28 \, \text{mm}$$

that is, a 1.41 times larger tolerance than the specified one is thus obtained.

In the case of a rectangular tolerance zone, where the dimensional tolerances are indicated with l_1 and l_2, the maximum allowable displacement is obviously equal to:

$$\sqrt{l_1^2 + l_2^2}$$

and is therefore equal to 0.45 mm for the case shown in Fig. 9.4b.

Coordinate positioning offers the following advantages:

(a) it is a simple and easy to understand system, and it is frequently used;
(b) it allows controls to be made with ordinary equipment, without the need of functional gauges or other more sophisticated systems.

However, it also introduces the following disadvantages:

(a) as already seen, a square or rectangular tolerance zone allows a variation of the greatest position of the indicated error; it is therefore necessary to specify, in the drawings, a restricted tolerance of 70% of that which would be functionally acceptable;
(b) it is possible to have an increase in the error when a dimensioning is used in series (accumulation of tolerances);
(c) there is a risk of functional components being discarded.

With a cylindrical zone positioning tolerance (assuming a tolerance value equal to the maximum allowable error in the coordinate system), a 0.28 mm diameter cylinder is obtained equal to the diagonal of the square zone (Fig. 9.5), with the advantage of obtaining 57% of additional tolerance.

Moreover, workpieces that would otherwise be discarded can be accepted, and this represents a significant manufacturing advantage.

However, the greatest advantage of the geometrical dimensioning tolerance is obtained through the use of the positioning control together with the Maximum Material requirement (MMR). In fact, if a maximum material modifier is inserted after the tolerance value (Fig. 9.6), an even greater increase in the tolerance zone is obtained: a tolerance of 0.28 mm when the hole is at the maximum material condition (9.9 mm diameter) becomes 0.48 mm when the hole is at the least material condition (10.1 mm diameter). A tolerance "bonus" (the portion of the tolerance zone that falls

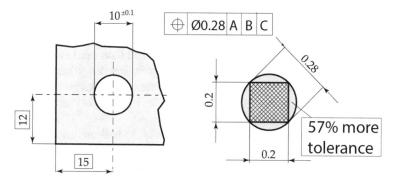

Fig. 9.5 The tolerance zone is cylindrical, with a bonus of 57% for the same tolerance as the square zone

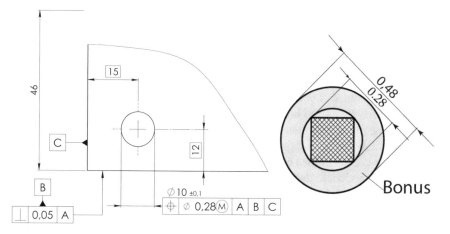

Fig. 9.6 Advantage of the application of the Maximum Material requirement

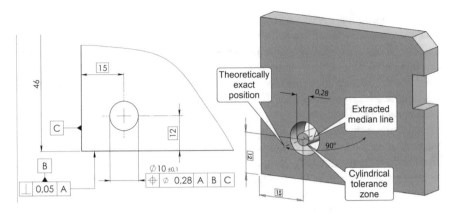

Fig. 9.7 Cylindrical tolerance zone of a position tolerance. It is advisable to indicate the surface that is theoretically perpendicular to the hole axis as the primary datum in order to control the perpendicularity deviation

between two cylinders with diameters of 0.28 mm and 0.48 mm) is thus obtained, and this leads to a further reduction in the number of discarded workpieces.

Another advantage of the use of geometrical tolerances is that it prevents unwanted tolerance accumulation over a dimensioning of holes in series, in that each tolerance is related to its own exact theoretical position.

It is advisable, in the case of position tolerances, to indicate the surface that is theoretically perpendicular to the hole axis as the primary datum in order to control the perpendicularity deviation. The extracted, or derived, median line must remain within a 0.28 mm diameter cylinder, the axis of which is perpendicular to datum A, and positioned in a theoretically exact way with respect to datums B and C. The cylindrical three-dimensional tolerance zone is of the same length as the thickness of the workpiece (Fig. 9.7).

9.2.1 Position Tolerance Applied to Median Surfaces

A position tolerance is generally used not only to control the position of features of size, such as shafts and holes, but also that of slots and tenons with a symmetry median plane that can be mated.

The position tolerance shown in Fig. 9.8 is used to control the position of a slot: the tolerance zone is limited by two parallel planes 0.2 mm apart, and symmetrically arranged with respect to the median plane of the feature whose position is fixed with a theoretically exact dimension with respect to datum B (Fig. 9.9).

The median plane of 8 external features, which are arranged radially, is instead controlled in Fig. 9.10. In this case, the extracted median surface should fall within

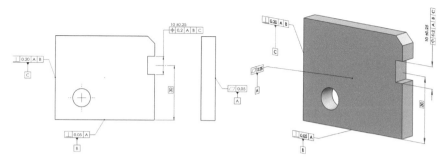

Fig. 9.8 Use of a position tolerance to control the location of a slot

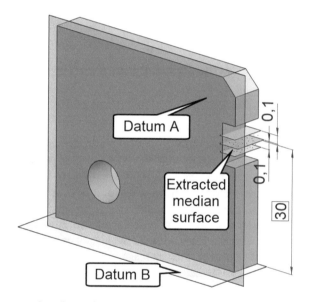

Fig. 9.9 The extracted median surface should fall between two parallel planes 0.1 apart, which are symmetrically arranged about the theoretically exact position of the median plane, with respect to datum planes A and B

two parallel planes 0.05 mm apart and arranged symmetrically around the theoretically exact position of the median plane, which is positioned with respect to datum A.

When the modifier SZ (Separate Zone) is used, the angle between the 8 tolerance zones is not restrained. On the other hand, the symbol CZ (Combined Zone) allows the tolerance zone to be blocked at intervals of 45°.

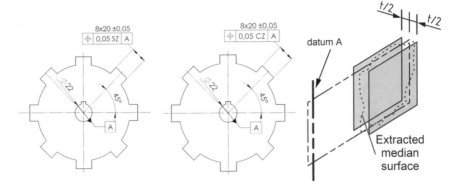

Fig. 9.10 Control of the median plane of 8 external features arranged radially. The two alternative symbols, SZ (Separate Zone) and CZ (Combined Zone), are used in the figure

9.2.2   Choice of the Modifiers for Position Tolerances

Three different methodologies exist for the control of the effect of a position tolerance:

1. **Regardless of feature size** (RFS in ASME), without any material condition modifier, as in the drawing of the Fig. 9.5. The specified geometric tolerance is independent of the actual size of the feature, according to ISO 8015.
2. **Axis or centreplane methodology**, in which an MMR or LMR requirement is used. In this case, an extracted median line or a plane of a feature of size should fall within the tolerance zone. The specified tolerance value applies only where the feature of size is at the MMC. Where the size departs from the MMC, the positional tolerance increases. This increase in the positional tolerance is equal to the difference between the actual size and the specified MMC tolerance value.
3. **Surface methodology**, in which a theoretical boundary (MMVC or LMVC) controls the positioning of the surface of a feature of size. In this case, *the indication of a modifier is obligatory*. While maintaining the specified size limits of the feature, no element of the surface should violate this theoretical boundary.

The interpretation of the position control of an axis and the surface of a hole is shown in Fig. 9.11. The most unfavourable condition is obtained by subtracting the position tolerance (0.2 mm) from the maximum material size (30 mm), thus a maximum material virtual size (MMVS) of 29.8 mm is obtained.

In the case of the *axis methodology*, when the hole is produced at minimum material condition (diameter of 30.3 mm), the position tolerance increase to a value of 0.5 mm (bonus). In this way, a dynamic and functional relationship is expressed between the dimension of the feature that is subject to a dimensional tolerance and the position tolerance, which means that wider holes and shafts of smaller diameter may lead to larger position errors.

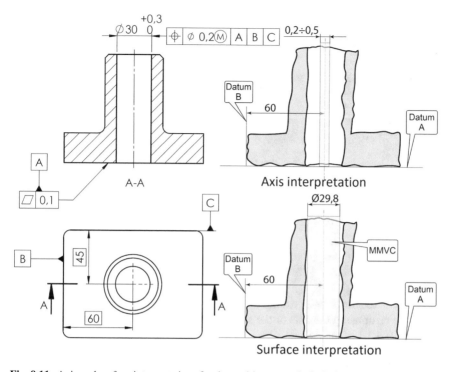

Fig. 9.11 Axis and surface interpretations for the position control of a hole

In the case of the *surface methodology*, the MMVC defines the theoretical mating boundary that should not be violated, whose position is controlled by the position tolerance, and which is a constant quantity. The axis of the cylindrical tolerance zone is fixed by theoretically exact dimensions with respect to datums A, B and C. The advantage of this specification is that it offers the possibility of utilising gauge pins for the control of the position of the holes.

The choice of modifiers (or of using no modifier), constitutes one of the most critical aspects for position tolerances. Table 9.2 summarises the choice criteria of the modifiers in function of the application modalities and the specific advantages that can be obtained.

Table 9.2 Selecting modifiers for position tolerances

Modifier	Uses	Examples	Notes
Ⓜ	Assembly, location	Boundary protection, fastener joints	Cost reduction, additional tolerances, functional gauging
Ⓛ	Minimum distance	Minimum wall thickness, minimum feature distance	Additional tolerances, variable gauging
RFS, no modifier	Symmetry, centring	Centring reference pin	Variable gauging

Fig. 9.12 According to the ASME Y14.5:2018 standard, if the size requirements are satisfied and MMVC is not violated, the feature is acceptable, even if the axis of the associated feature is outside the positional tolerance zone

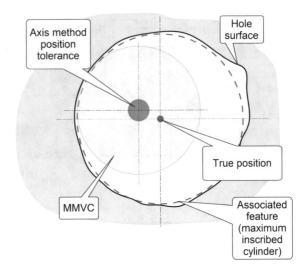

Figure 9.12 shows, as an example, a typical case of the interpretation of a position tolerance error using the axis or the surface method. According to the rule outlined in Sect. 10.3.3.1 of the ASME Y14.5:2018 standard, the surface method should take precedence over the axis method. Instead, if the size requirements are satisfied and the virtual condition is not violated, the feature is acceptable, even when the axis of the associated feature is *outside* the positional tolerance zone.

9.2.2.1 Effects of Specifying the MMR Requirement

Functional design is founded on a new way of conceiving a drawing, which is based on the dimensioning of workpieces according to their way of functioning; this is achieved with the advantage of improving communication, reducing controversies and above all introducing higher tolerances according to the philosophy: "making the largest possible tolerance allowed with the function available".

The following example shows how it is possible to increase the value of a tolerance, without compromising the functional conditions, considering the effects of the gain in tolerance due to the application of a maximum material requirement to either the feature itself (*bonus*), or to the datum (*shift*).

Let us consider what is shown in Fig. 9.13, where plate "L" is mounted with the use of a screw and a reference pin. These functional conditions lead to the drawing of the plate shown in the same figure. The 11 mm hole is positioned with respect to primary datum A (to control the orientation) and with respect to secondary datum D (axis of a 16 mm the hole). Since the secondary datum is a feature of size, it is possible to reference the datum feature according to the MMR requirement. In this case, if the pin hole is produced at the least material condition (as wide as possible,

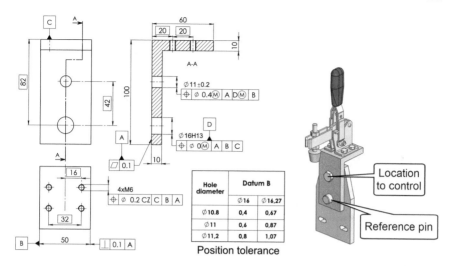

Hole diameter	Datum B	
	Ø16	Ø16,27
Ø10.8	0,4	0,67
Ø11	0,6	0,87
Ø11,2	0,8	1,07

Position tolerance

Fig. 9.13 Datum A is considered a primary datum for the dimensioning of the L-shaped plate, and it controls the orientation of the 11 mm hole, while secondary datum D is a feature of size, and it is therefore possible to apply a modifier M, which allows a further bonus, called shift, to be obtained

that is, 16.27 mm), the position tolerance increases up to a value of 1.07 mm, as shown in the table.

However, it is necessary to pay particular attention to this type of indication since this additional tolerance is similar to a bonus, *but it is not a bonus*, and it is only available under certain conditions.

In short, the application of a maximum material modifier to a datum does not increase the tolerance zone, but simply allows *an allowable movement of the datum feature (which is called "shift") with respect to the functional datum.*

Let us consider the coupling shown in Fig. 9.14, in which the mating between the plate takes place with 6 fixed fasteners. The two functional datums are constituted

Fig. 9.14 Mating between two plates with fixed fasteners

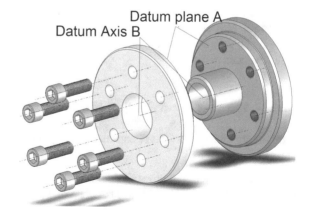

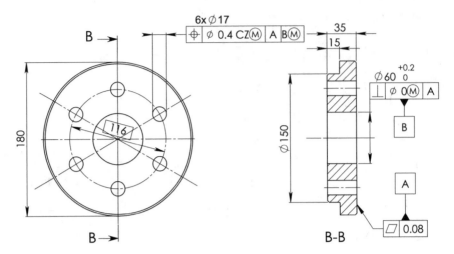

Fig. 9.15 Functional dimensioning of a flange: datum A is considered a primary datum, and it controls the orientation of the 6 holes, while secondary datum B is an axis, and it is therefore possible to apply a maximum material requirement

by the planar boundary surface (primary datum A) and the centring axis (secondary datum B). A third datum is not necessary, because of the symmetry of the holes.

These functional conditions lead to the dimensioning of the first flange, as shown in Fig. 9.15. The six 17 mm diameter holes are located with respect to primary datum A (to control the orientation) and with respect to secondary datum B (60 mm hole axis). As the latter datum is a feature of size, it is possible to apply the maximum material requirement (MMR).

In this case, if the central hole is produced at the least material condition (as wide as possible, that is, 60.2 mm, with perfect perpendicularity with respect to the primary datum), the position tolerance of the 6 holes does not increase. In fact, in these conditions, the set of the 6 holes can only shift in any direction by 0.2 mm with respect to the functional datums (Fig. 9.16).

It is thus possible to recover components that would otherwise be discarded during the control. However, it is advisable to only use this type of indication when the verification can only be conducted with functional gauges or with coordinate machines with advanced software.

9.2.2.2   Effects of Specifying the LMR Requirement

A minimum material modifier (least material requirement, LMR) is generally utilised in three applications:

(a)   to control the minimum wall thickness;
(b)   to ensure a minimum distance between one feature and another;
(c)   to control a well-defined angular position.

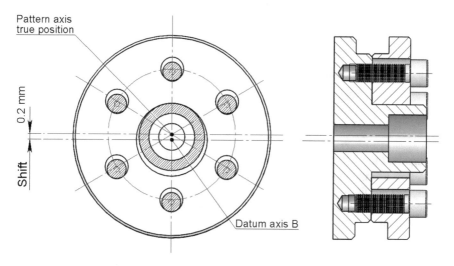

Fig. 9.16   The central hole is at the least material condition (60.2 mm) with perfect perpendicularity. The modifier M on the indication of the datum B does not increase the position tolerance of the 6 holes, but it does allow a shift of the 6 holes by 0.2 mm, with respect to the functional datum, in any direction

In the case of Fig. 9.17, the LMR requirement is utilised when the greatest accuracy of the radial slot angular position is required in an axisymmetric part; the desired distance between the border of the slot and the theoretical position of the median plane of the same slot is thus maintained. In fact, for the least material condition, we obtain:

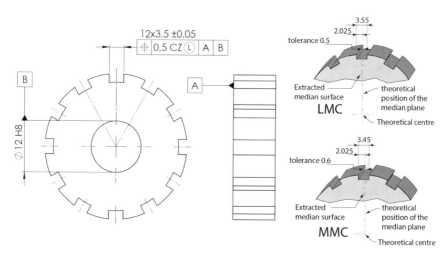

Fig. 9.17   Position tolerance with LMR applied to a series of radial slots. The distance between the face of a slot and the median plane of the datum is always kept constant (2.025 mm)

$$\frac{3.55}{2} + \frac{0.5}{2} = 2.025$$

while, for the maximum material condition (the tolerance increases to 0.6 mm), we obtain:

$$\frac{3.45}{2} + \frac{0.6}{2} = 2.025$$

Finally, Fig. 9.18 shows the utilisation of the least material requirement for the control of a wall thickness. The position tolerance is more limited (0.25 mm) when the hole is produced at the minimum material condition (4.2 mm, larger hole), while it increases up to a value of 0.5 mm at the maximum material condition (3.95 mm, smaller hole).

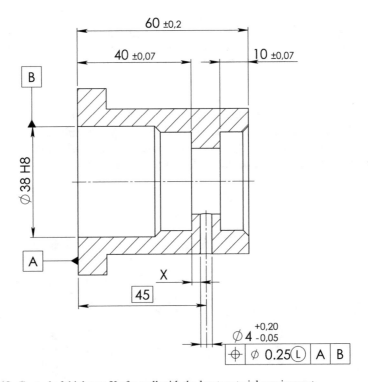

Fig. 9.18 Control of thickness X of a wall with the least material requirement

9.2.3 Pattern Location

The ISO 5458 standard establishes rules that are complementary to those of ISO 1101 to apply to **pattern specifications** and defines rules on how to combine individual specifications, for geometrical specifications e.g. using the position, symmetry, line and surface profile symbols as well the flatness one (in the case where the toleranced features are nominally coplanar). A pattern of features of size may require multiple levels of positional control, since it may require a larger tolerance for the datum system, but a smaller tolerance within the pattern.

According to the aforementioned standard, a pattern specification consists of both a set of more than one geometrical feature and a tolerance zone pattern. The set of tolerance zones in the tolerance zone pattern have *internal constraints*, which are defined by means of implicit or explicit TEDs. If necessary, *external constraints* to a tolerance zone pattern can be defined by referencing a datum system, with implicit or explicit TEDs.

The specification in Fig. 9.19 is not a pattern specification (SZ modifier). The toleranced feature is a collection of two extracted median lines. Each individual tolerance zone is considered independently (SZ modifier) and *does not constitute a tolerance zone pattern*. Each tolerance zone is a cylindrical zone with a diameter of 0.2 mm, where the axis is externally constrained in orientation to be parallel to datum A (implicit TED of 0°) and in location at a distance 25 mm (explicit TED) from datum A. The tolerance zones of the two toleranced features are independent of each other and are not constrained between one another. The distance of 50 mm (25 + 25) is not considered *as an internal constraint* between the tolerance zones (SZ modifier).

The specification in Fig. 9.20 is *a pattern specification* (CZ modifier). The toleranced feature is the collection of two extracted median lines. The tolerance zone is a tolerance zone pattern (CZ modifier) composed of two 0.2 mm diameter cylindrical zones, where their axes are internally (CZ modifier) constrained to be parallel in

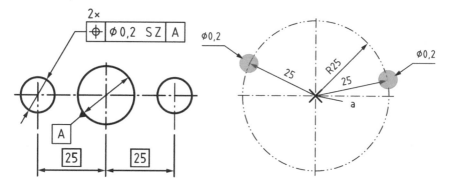

Fig. 9.19 Location of two independent tolerance zones from datum A

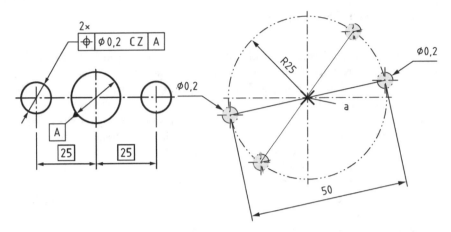

Fig. 9.20 Pattern specification. Two tolerance zones constrained between one another 50 mm apart and located at a distance of 25 mm from datum A

orientation and 50 mm apart in location. Moreover, the tolerance zones are externally constrained in location at a distance of 25 mm from datum A.

9.2.3.1 Indication of a Multiple Indicator Pattern Specification

In order to create a multiple indicator pattern specification, the *SIM* (simultaneous requirement) *modifier* should be indicated in the adjacent indication area of each related geometrical specification, with the option of following it with an identification number without a space.

The use of the SIM modifier transforms a set of more than one geometrical specification into a pattern specification. The tolerance zones of all the specifications are thus locked together with location and orientation constraints.

Two simultaneous requirements, defined by SIM1 and SIM2 indications, are shown in Fig. 9.21, and each simultaneous requirement should be considered individually. The two specifications linked together with the SIM1 indication both use a CZ modifier to create a tolerance zone pattern. The tolerance zone pattern is composed of two combined tolerance zone patterns:

- the first is a set of three tolerance zones, consisting of two parallel planes 0.1 mm apart, for the three extracted median surfaces of the 14 mm slots;
- the second is a set of three tolerance zones, consisting of two parallel planes 0.15 mm apart, for the three extracted median surfaces of the 13 mm slots.

The SIM1 modifier locks the two tolerance zone patterns together into a combined tolerance zone pattern of six (3x + 3x) tolerance zones.

The two specifications linked together with the SIM2 indication both use a CZ modifier to create a tolerance zone pattern. One of the specifications is a pattern

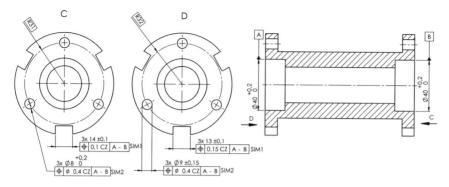

Fig. 9.21 Example of two separate simultaneous requirements applied to different pattern specifications

of three Ø0.4 tolerance zones for the three extracted median lines of the Ø8 holes, and the other is a tolerance zone pattern of three Ø0.4 tolerance zones for the three extracted median lines of the Ø9 holes. The SIM2 modifier locks the two tolerance zone patterns together into a combined tolerance zone pattern of six (3x + 3x) cylindrical tolerance zones.

9.2.3.2 Indication of a Multi-Level Pattern Specification

The symbols in Table 9.3 are used to describe a multi-level single indicator pattern specification utilised to control n *identical features* grouped in k *identical groups*. A set of k identical groups, each consisting of n single identical features, should be indicated, using the notation kx and nx, to create a *multi-level single indicator pattern specification*; kx and nx should be followed by a space and an identifier letter or symbol, in order to avoid ambiguities, with a slash as a separator and a space on both sides of the slash (e.g. *4x/2 × or 4 × A/2 × B*).

The identification letter can be used to establish a link with individual integral features, or with a group of integral features. The groups and the features should be indicated with a leader line and with a capital letter.

If the first element of the sequence is SZ and the following element is SZ, then the tolerance zone patterns are separate and independent of each other (Fig. 9.22). In short, there are kx n independent tolerance zone patterns and the *specification does not define a pattern specification*.

	Symbol	Interpretation
Table 9.3 Symbols used to describe a multi-level single indicator pattern specification	k	Number of identical groups
	n	Number of identical features

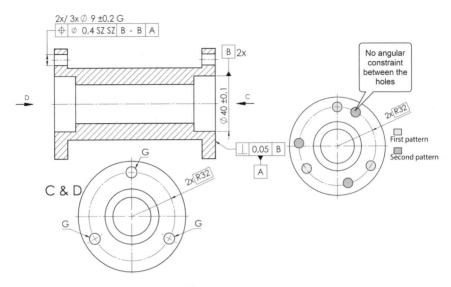

Fig. 9.22 Six independent pattern specifications

If the first element of the sequence is SZ and the following element is CZ, then there are *k independent tolerance zone patterns* (SZ), composed of n *individual tolerance zones locked together with orientation and location constraints* (CZ).

If all the elements of the sequence are CZ, then there is *one tolerance zone pattern*, and the specification consists of one combined zone (tolerance zone pattern), applied to a set of k x n geometrical features.

The first specification (SZ CZ) in Fig. 9.23 controls two independent pattern specifications. The tolerance features are sets of extracted median lines and the two tolerance zone patterns are independent of each other (no angular constraint between them). Each tolerance zone pattern is a combined zone of three 0.2 mm diameter cylinders, constrained between each other in orientation (parallelism) and in location (equidistanced angularly on a 32 mm diameter cylinder).

The second specification defines a two-level pattern (CZ CZ: a tolerance zone pattern of two tolerance zone patterns) applied to six (2 × 3) extracted median lines.

Two tolerance zone patterns are defined with the last CZ of the sequence (CZ CZ). Each tolerance zone pattern is a combined zone of three 0.5 mm diameter cylinders, constrained between each other in orientation (parallelism) and in location (equidistanced angularly on a 32 mm diameter cylinder).

The two tolerance zone patterns are not independent, as they are *constrained to each other in orientation* (parallelism) and *in location* (the axes of each tolerance zone pattern are coaxial, 0 mm, and the tolerance zone patterns are rotationally locked at 0°).

When used to identify a group of features, the group may be indicated on a drawing by surrounding the features with a long-dashed double-dotted narrow line (05.1 line type according to ISO 128-24, see Fig. 9.24).

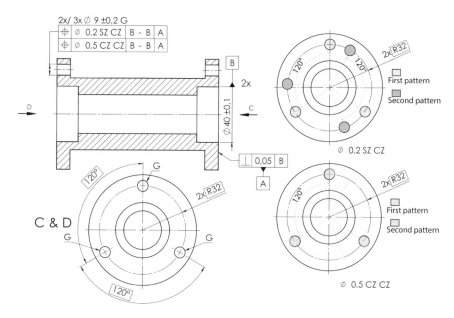

Fig. 9.23 Example of indication of a two-level single indicator pattern specification

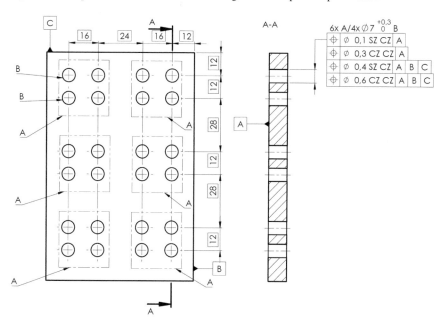

Fig. 9.24 Example of an indication of a multi-level single indicator pattern specification

The first specification (SZ CZ with datum A) in Fig. 9.24 controls six independent pattern specifications. The tolerance zone pattern (combined zone) for each pattern specification is composed of four 0.1 mm diameter cylindrical zones with an *orientation* constraint (parallel to each other and perpendicular to datum A) and with a *location* constraint between one another (16 mm apart in a horizontal direction and 12 mm apart in a vertical direction).

The second specification (CZ CZ with datum A) controls six dependent pattern specifications and results in only one pattern specification. The tolerance zone pattern (combined zone) is composed of twenty-four 0.3 mm diameter cylindrical zones with an *orientation* constraint (parallel to each other and perpendicular to datum A) and with a *location* constraint between each other (16 mm [24 mm between the groups] apart in a horizontal direction and 12 mm [28 mm between the groups] apart in a vertical direction).

The third specification (SZ CZ with datums A, B and C) defines six independent pattern specifications constrained in location by datum B and datum C. The tolerance zone pattern (combined zone) of each pattern specification is composed of four 0.4 mm diameter cylindrical zones with an *orientation* constraint (parallel to each other and perpendicular to datum A) and with a *location* constraint between each other (16 mm apart in a horizontal direction and 12 mm apart in a vertical direction and constrained by datums B and C at a distance of 12 mm).

The fourth specification (CZ CZ with datum system A, B and C) controls six dependent pattern specifications, constrained in location from datum B, resulting in only one pattern specification. The tolerance zone pattern (combined zone) is composed of twenty-four 0.6 mm diameter cylindrical zones with an orientation constraint (parallel to each other and perpendicular to datum A) and with a location constraint between one another (16 mm [24 mm between the groups] apart in a horizontal direction, 12 mm [28 mm between the groups] apart in a vertical direction and constrained by datums B and C at a distance of 12 mm).

An alternative specification indicated by a CZ CZ modifier in the tolerance section, is a specification (with the same meaning) applied to *one tolerance zone pattern* consisting of m(=k × n) geometrical features, indicated by one CZ modifier in the tolerance section, as illustrated in Fig. 9.25.

9.2.4   Calculation of a Geometrical Position Tolerance

When geometrical position tolerances are applied, the value of the tolerance is calculated from the mating conditions, that is, from the maximum and least dimensions allowed for the features that have to be mated.

Figure 9.26 shows the case of a mating of a plate with four holes and a centring connection hole, attained by means of bolted joints (that is, a bolt and a nut) of a nominal diameter M6 to a second plate with a central pin; the problem arises of how to calculate the position tolerances of the plate with clearance holes for

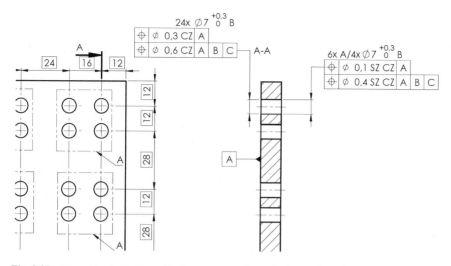

Fig. 9.25 Alternative indication with the same meaning as in the previous figure

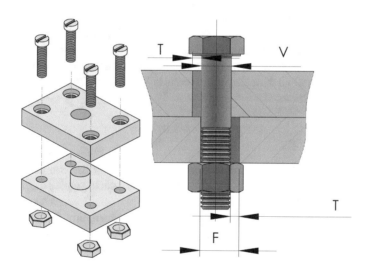

Fig. 9.26 Example of a Floating Fastener Assembly

the fasteners. Therefore, the functional condition requires the use of the MMVC boundary methodology, that is, with the maximum material requirement.

The floating fastener[1] formula is:

$$T = F_{MMC} - V$$

---

[1]Par. 10.3.3.2 ASME Y14.5:2018.

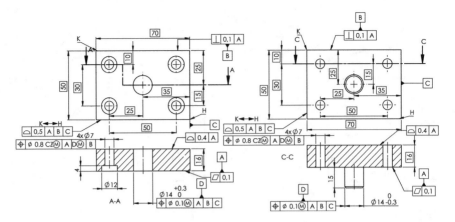

Fig. 9.27 Drawing of the two plates joined by bolts with the calculated position tolerance

where F_{MMC} is the diameter of the clearance hole at maximum material conditions and V is the size of the fastener (the relative tolerance is neglected).

Let us assume that a hole diameter of 7 mm has been chosen for both plates (clearance hole for fasteners, ISO 273) and that deviations of ± 0.2 mm have been assumed for the holes (foreseen according to ISO 2768-m for general dimensional tolerances). The diameter of the hole at the maximum material condition is FMMC = 6.8 mm and the nominal diameter of the screw is V = 6 mm. Therefore, the position tolerance of the holes is:

$$T = 6.8 - 6 = 0.8\,\text{mm}$$

which applies to each part of the assembly. The mating of the two plates, for the most critical conditions, is also shown in Fig. 9.26, which graphically shows the correctness of the adopted formula.

Figure 9.27 illustrates the proposed functional dimensioning of the two plates with the calculated position tolerances.

Figure 9.28 instead indicates the joining of the same two plates when fixed fasteners are used; in this case, the holes in one component of the assembly are threaded holes. The fixed fastener[2] formula is:

$$T = \frac{F_{MMC} - V}{2} = \frac{6.8 - 6}{2} = 0.4\,\text{mm}$$

The total tolerance may be divided equally between the two plates; the formula can be demonstrated considering the mating with fixed fasteners shown in the same figure. The position tolerance of plate A (T_1) and plate B (T_2) can also be calculated with the following formula:

---

[2]One of the parts that has to be assembled has restrained fasteners.

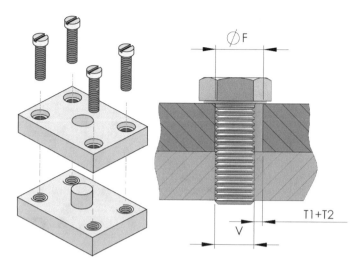

Fig. 9.28 Mating of the two plates with fixed fasteners: calculation of the position tolerances

$$T_1 + T_2 = F_{MMC} - V$$

The dimensioning proposal of the two plates is shown in Fig. 9.29 (the omission of a maximum material modifier on the threaded hole should be noted).

Unfortunately, the preceding formulas do not provide sufficient clearance for the fixed fastener case when threaded holes are out of square. In fact, the inclination of a fixed fastener is governed by the inclination of the threaded hole in which it is assembled, and this effect could cause fasteners, such as screws, studs or pins, to

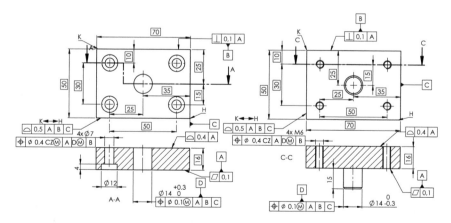

Fig. 9.29 Drawing of the two plates connected by means of fixed fasteners with the calculated position tolerance. It should be noted that a maximum material requirement is not used to control the position of the M6 threaded holes

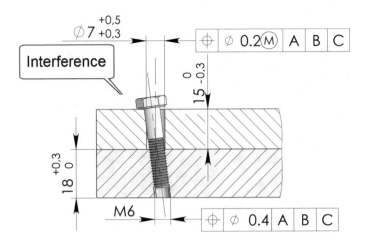

Fig. 9.30 This figure points out that the inclination of the axis of the screw is governed by that of the threaded hole (the inclination is within the position tolerance limits of 0.4 mm). In order to avoid any interference, the minimum diameter of 7.3 mm of the hole is chosen according to the thickness of the two components and the position tolerances

interfere with the mating parts (Fig. 9.30). In order to take into consideration this condition, the projected tolerance zone method of positional tolerancing should be applied to threaded holes (as explained in the next section).

When the projected tolerance zone system is not used, it is necessary to select a positional tolerance and hole clearance combination that compensates for the allowable tilting of the axis of the fixed fastener feature. The following formula may be used[3]:

$$H = V + T_1 + T_2\left(1 + \frac{2S}{P}\right)$$

where S is the maximum thickness of the plate with the clearance hole, P is the minimum thickness of the plate with the threaded hole, and T_1 and T_2 are the positional tolerance of the clearance hole and the positional tolerance of the tapped hole, respectively.

When the formula for the dimensions in Fig. 9.30 is applied, we obtain:

$$H = 6 + 0.2 + 0.4\left(1 + \frac{30}{18}\right) \cong 7.3mm$$

Figures 9.31 and 9.32 show simple practical examples of the calculation of position tolerances for the joining with floating and fixed fasteners; as can be seen, it is possible to divide the tolerance between the various features in both a uniform and

---

[3] Appendix B-5 of ASME Y14.5:2018.

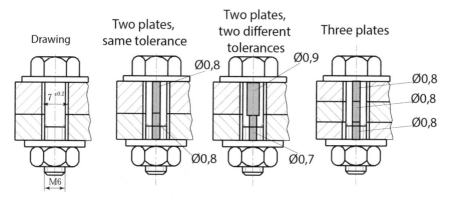

Fig. 9.31 Practical examples for the calculation of position tolerances for the joining with floating fasteners

Fig. 9.32 Practical formulas that can be used to calculate the position tolerances for connections with fixed fasteners

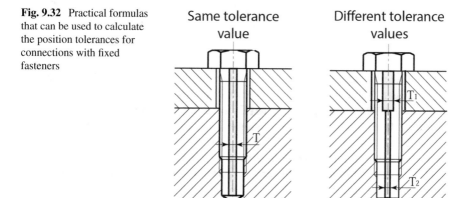

non-uniform way (for example, using higher tolerances on components that are more expensive to manufacture).

9.2.5 Projected Tolerance Zone

The application of this concept is recommended whenever a variation in perpendicularity of threaded or press-fit holes could cause fasteners, such as screws, studs or pins, to interfere with mating parts. Let us consider, for example, the fixed fastener shown in Fig. 9.33; the position tolerance indicated in the drawing, and calculated by

Fig. 9.33 The calculated
position tolerance is
respected in a joint with
fixed fasteners, but when the
assembly is carried out, an
interference arises in that the
inclination of the axis of the
screw is governed by that of
the hole

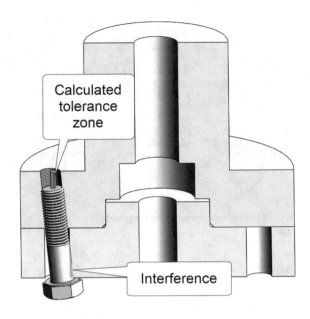

means of the previously seen formulas, is complied with, but when the assembly is
carried out, an interference condition near the head of the fastener occurs. In order to
avoid this problem, it is a good idea to utilise the **projected tolerance zone** concept,
according to which the position tolerance is not established on the feature itself,
but on its external projection. In practice, it is necessary to imagine projecting the
tolerance zone outside the component for a certain length.

In this case, the toleranced feature is a portion of the extended feature, which is
an associated feature constructed from the integral feature. The default association
criterion for the extended feature is *a minimised maximum distance between the
indicated integral feature and the associated feature* with the additional constraint
of external contact with the material.

As can be deduced from Fig. 9.34, tolerance zone T is outside the threaded hole,
whose position error may even be larger that T.

Such a projected zone should be indicated with a ℗ symbol placed inside the
tolerance indicator, and the length of the projected external zone should then be
indicated, as shown in Fig. 9.35. As an alternative, it is possible to directly indicate
the projected toleranced feature length by means of a *"virtual" integral feature* that
represents the portion of the extended feature that should be considered. The virtual
feature should be indicated in a clearer way with a long-dashed, double-dotted narrow
line (05.1 line type according to ISO 128-24), as shown, as an alternative, in Fig. 9.35.

The extension of the projected tolerance zone ℗ outside the threaded hole is a
minimum value, and it also represents *the maximum thickness of the workpiece that
has to be mated*. Whenever one does not want to apply this concept, it is preferable
to *combine a perpendicularity tolerance with the position requirement*.

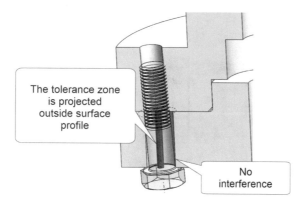

Fig. 9.34 The tolerance zone is projected outside the threaded hole, in order to avoid any interference

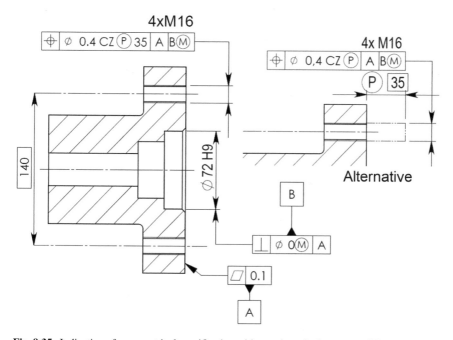

Fig. 9.35 Indication of a geometrical specification with a projected tolerance modifier using direct and indirect indication of the length of the projected toleranced feature in the tolerance indicator

By default, the origin of the projected feature should be at the location of the reference plane, and the end corresponds to the shift in the length of the projected feature from its origin in the direction outside the material. If the origin of the projected feature is displaced from the reference surface by an offset, this should be specified by means of a theoretically exact dimension (Fig. 9.36). As an alternative indication, the first value after the modifier should indicate the distance from the

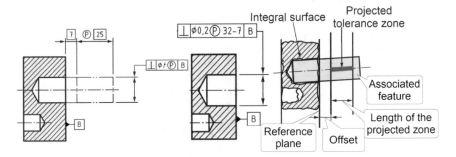

Fig. 9.36 Interpretation of an indirect indication of a projected tolerance with an offset

farthest limit of the extended feature and the second value (offset value), which is preceded by a minus sign, should indicate the distance from the nearest limit of the extended feature (the length of the extended feature is the difference between these two values).

A possible control method of the projected tolerance with a coordinate measuring machine that controls the positioning of threaded cylindrical features, with a projection equal to the length of the projected tolerance zone, screwed onto holes that have to be controlled, is shown in Fig. 9.37.

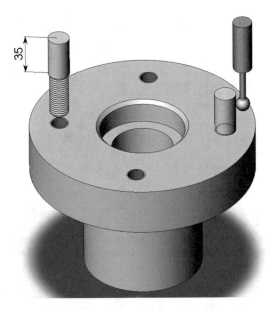

Fig. 9.37 Inspection of the projected tolerance zone: a coordinate measurement machine controls the position and orientation error of the functional gauges, which is of a length that is equal to the projected tolerance. The gauges are screwed onto the workpiece

9.2.5.1 Indication of a Datum with a ⓟ Modifier

When the ⓟ modifier is placed in the tolerance frame after a letter indicating a datum established from a feature of size, then the datum feature should be established by fitting an associated feature of the projected length to the extension of the real feature and not to the real integral feature itself. The ⓟ modifier can be applied to a secondary or tertiary datum, but has no effect when it is applied to a primary datum. An example of the application of modifier ⓟ to a secondary datum is shown in Fig. 9.38.

9.2.6 Position Tolerances in the ASME Standards

There is a substantial difference in the specification of a position tolerance in the ISO and ASME standards. The ASME standards specify, without any shadow of doubt, that *the position tolerance symbol should only be utilised for a "feature of size"*, while the ISO standards allow it to be used to position planar surfaces (Fig. 9.39). In order to avoid specification errors (for example, the erroneous use of maximum material modifiers), it is advisable to always follow the ASME standard indications.

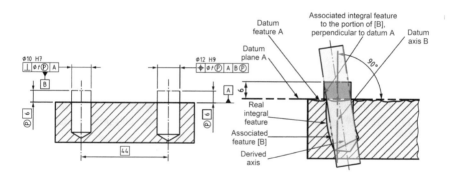

Fig. 9.38 Example of the application of modifier ⓟ to a secondary datum

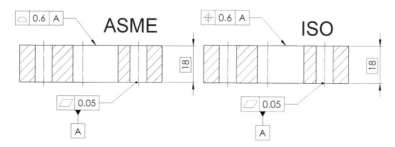

Fig. 9.39 The use of the position symbol for planar surfaces is forbidden in the ASME standard

The ISO and ASME standards use two different approaches to control the location of a feature of size: the ISO standards, in order to locate a feature of size, define the concept of extracted median line or median surface. Instead, in the ASME standards, the axis or median plane is used to control the location of a feature of size.

9.2.6.1 Composite Positional Tolerances in ASME Y14.5

There are many practical applications in which the positioning of the holes of a component, with respect to the datums, is less important than the accuracy of the position of each single hole within the pattern of holes.

Let us consider the drawing of the plate shown in Fig. 9.40 in which, as a design requirement, the joint with bolts is specified and, at the same time, a great precision in the positioning of the holes in the plate, with respect to the datums, is not necessary.

In this case, the position tolerance is specified by a dual tolerance indicator that has one position symbol, which is specified with two tolerance information lines in order to indicate a position tolerance for both the control of the location of the features with respect to one another (lower segment of a composite position tolerance frame) and for the location of the pattern of the holes with respect to the specified datums (upper segment).

In order to obtain a correct interpretation of the drawing, it is necessary to take into due consideration that each indication must be respected independently, that is, the lower frame should indicate that the derived median line of each of the four holes, in the maximum material condition, should fall within a 0.1 mm diameter tolerance zone. This constitutes **a hole-hole position relationship** of the set.

The upper frame refers to the pattern of holes and, in practice, specifies the orientation and positioning of the pattern of holes: in this case, the axis of each hole should fall within a 0.3 mm diameter cylindrical tolerance. The position tolerance

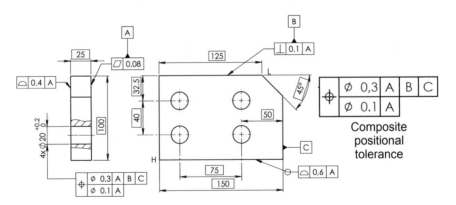

Fig. 9.40 A composite positional tolerance with multiple segments may be used for the case in which the positioning of each single hole within a pattern of holes is important, while the positioning of the holes with respect to the datums does not need to be very precise

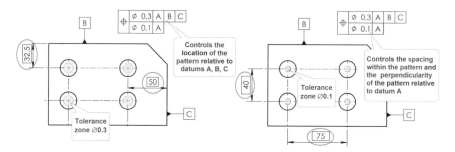

Fig. 9.41 Interpretation of the upper and lower frames of a composite tolerance. The circled dimensions indicate the distances controlled by each segment of the composite frame

zones are placed in their theoretically exact positions with respect to three datums, A, B and C. This constitutes **a position relationship** of a pattern of holes. The tolerance zone in the lower frame should always be smaller than the tolerance zone in the upper frame. In composite tolerances, the feature-feature relationship (lower frame) is defined with the term *Feature-Relating Tolerance Zone Framework*(**FRTZF**, which is pronounced "*Fritz*"), while, for position tolerances of the set of features (upper frame), the *Pattern-Locating Tolerance Zone Framework* term (**PLTZF**, pronounced "*Plahtz*") is used.

Figure 9.41 clarifies the interpretation of the position of the pattern of the four holes with respect to the datums. Therefore, two levels of control of the pattern of holes exist: the position tolerance zones of the upper frame are placed in their theoretically exact position with respect to three datums, A, B and C. The tolerance zones in the lower frame control the position and orientation errors of a hole with respect to another hole and to the perpendicularly to the specified datum, A. The pattern of holes may therefore rotate and move within the tolerance boundaries of the upper segment, but should remain perpendicular to referenced datum A.

Many practical applications exist in which the positioning of the holes of a plate, with respect to the boundary, is less important than the orientation of the pattern of holes. A typical example is the plate with the assembly shown in Fig. 9.42, whose position, with respect to the edges, is less problematic than the orientation of the holes, which would produce an aesthetically negative result. The upper frame has a wide tolerance in order to specify the position of the holes with respect to the borders. The lower frame has a narrow tolerance, and it not only specifies the spacing between the holes, but also controls the orientation of the two holes with respect to datum A (perpendicularity) and datum B (parallelism, Fig. 9.43).

As noted in the previous paragraphs, in ISO standard 5458 a pattern of features of size may have multiple levels of positional control as a result of the use of appropriate symbols. Since the ISO standard *no longer uses composite tolerance*, it is possible to specify the tolerance of the multiple level positions with the symbols shown in Fig. 9.44.

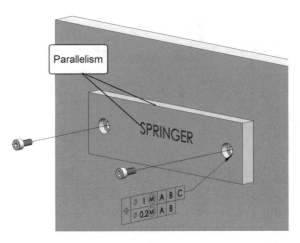

Fig. 9.42 In this example, the position with respect to the edges of the plate is less problematic than the orientation of the pattern of the holes, which would produce an aesthetically negative result. A composite position tolerance is therefore very useful

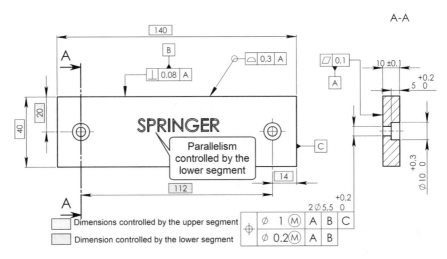

Fig. 9.43 The use of composite positioning. The upper frame has a large tolerance in order to specify the position of the holes with respect to the edges. The lower frame has a narrow tolerance in order to control the error over the distance between holes and, at the same time, it specifies the orientation of the two holes with respect to datum A (perpendicularity) and datum B (parallelism)

9.2.6.2 Control with Functional Gauges

As pointed out in the previous sections, the ISO standard is defined as "CMM Friendly", that is, the preferred control system is the coordinate measurement machine. The ASME standard is based on the idea of specifying the geometrically perfect zones within which the real surfaces should fall. This is often indicated as a

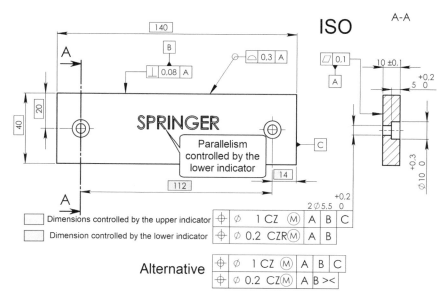

Fig. 9.44 Since the ISO standard no longer uses composite tolerance, it is possible to specify the tolerance of the multiple level positions with the symbols CZR and > <,

preference for "hard gauging", which means that it is possible to construct functional gauges that represent a physical representation of the tolerance zone.

A functional gauge basically represents the materialisation of the feature that has to be mated (worst case) according to the specifications indicated on the drawing. In short, a gauge is nothing more than a simulated physical datum feature that allows the relationships between geometrical and dimensional errors to be verified at the same time, and the effect of an increase in tolerance, due to a maximum material modifier applied to either the feature itself (bonus) or to the datums (called shift or MMB), to be foreseen. The tolerance on a gauge is generally about 10% of the tolerance that has to be controlled, under a temperature condition of 20° and humidity no higher than 45%. If a functional gauge is mated with the piece that has to be controlled, it is possible to be almost absolutely certain of the correct assembly with the mating counterpart.

A functional gauge is a gauge that is built to a fixed dimension (the virtual condition) of a part feature. A part must be able to fit into (or onto) the gauge. A functional gauge does not provide a dimensional measurement; it only indicates whether the part is compliant or not with the drawing specification. Since these control instruments have fixed dimensions, the additional tolerance (bonus) that is allowed, for example, for a hole produced at the limits of the maximum dimensions (or the dynamic "shift" of a datum subject to dimensional variations) is easily "captured" by the functional gauge. Moreover, functional gauges can easily be used by personnel with the minimum preparation in metrology, and they can significantly reduce the overall geometrical and dimensional verification times.

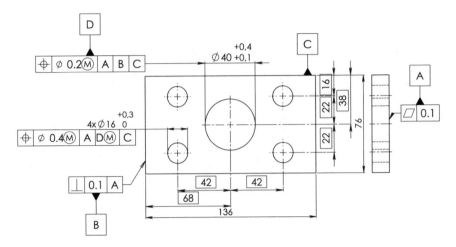

Fig. 9.45 Dimensioning of a plate (ASME) with the positioning of the central hole with respect to three functional datums, while the 4 holes are correlated with the 40 mm hole, considered as a datum and indicated with the letter D (with a maximum material modifier)

A drawing of a plate is shown in Fig. 9.45, where the central hole and the four 16 mm holes are positioned with respect to a 3 plane datum system, DRF, while the four 16 mm holes are correlated with the central hole, considered as a datum and indicated with the letter D. As datum D is a feature of size, it is possible to apply the position tolerance with a maximum material modifier applied to either the feature itself (bonus) or to the datum (shift). The plate is controlled with the help of two functional gauges, one for the control of the central hole (DRF system [A, B and C]) and the other for the four 16 mm holes (DRF system [A, D and C]).

As can be observed in Fig. 9.46, a functional gauge, constituted by three perpendicular surfaces that simulate the datums (A, B and C), is used to control the hole. The central pin of the gauge (whose correct insertion allows the workpiece to be accepted) has a diameter that corresponds to the virtual dimension, which means that it is necessary to subtract the geometrical tolerance from the maximum material dimensions of the hole: $40.1 - 0.2 = 39.9$ mm.

The second gauge is constituted by a simulator of datum A, a mobile plane used to simulate datum C and a fixed pin with the virtual dimensions of 39.9 mm (Fig. 9.47). The mobile datum only has the purpose of blocking the component, by stopping the rotation. The verification is carried out with 4 mobile pins at a virtual dimension of 16 mm $- 0.4$ mm $= 15.6$ mm.

In this way, when the diameter of datum D is larger than 40.1 mm, it is possible to translate (shift) the component in order to compensate for a position error of the pattern of 4 holes. The shift does not increase the position tolerance, but it does allow the pattern of 4 holes to be shifted in any direction with respect to the functional datums.

Functional gauges offer the following advantages:

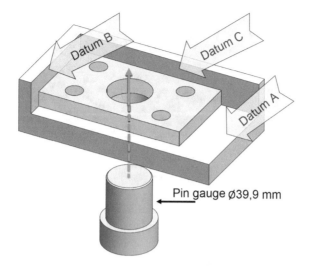

Fig. 9.46 Control of the central hole of a plate by means of a functional gauge. The central pin has a diameter that corresponds to the virtual dimension

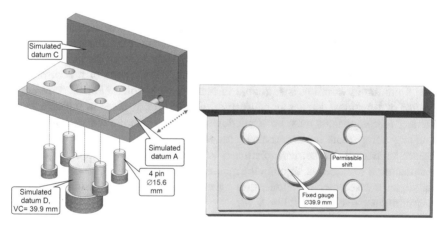

Fig. 9.47 The second gauge used to control the position error of the four 16 mm holes. When the datum has a maximum material modifier, the gauge is made up of a pin fixed at the virtual dimension of 39.9 mm. In this way, when the datum diameter D is larger than 40.1 mm, it is possible to translate (shift) the component to compensate for the position error of the pattern of 4 holes

(1) a workpiece can be controlled immediately and quickly;
(2) particular technical knowledge is necessary to use them;
(3) a workpiece that does not comply with the tolerance is never accepted;
(4) they represent the physical materialisation of the workpiece that has to be mated;
(5) they allow the effect of the bonus and of the virtual conditions to be understood.
     However, they also suffer from the following disadvantages:

Fig. 9.48 Control of the
position of a hole utilising a
coordinate measuring
machine (CMM)

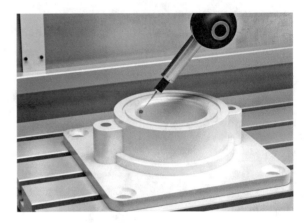

(a) they are expensive to build;
(b) even a minimum variation of the design of a workpiece renders them
     obsolete;
(c) they do not quantify the results of the control;
(d) they require a great design effort.

In conclusion, functional gauges are enormously costly, in design, building and
maintenance terms, and they require that a part of the tolerance that has to be verified is
sacrificed (usually about 10%) in order to supply the tolerance for the manufacturing
of the gauge itself. Moreover, they do not quantify the results of a control, and
even a minimum variation is sufficient to make them obsolete. For these reasons,
the use of functional gauges is generally limited to those cases in which a large
quantity of components must be verified in such a way that the reduced inspection
time may compensate for the elevated cost of producing the gauge. For this reason,
the verification of geometrical tolerances is carried out, for the great majority of
manufactured workpieces, through the manual elaboration of the data collected with
traditional measurement equipment, or by means of software, utilising coordinate
measuring machines (CMM, Fig. 9.48).

9.3 Concentricity and Coaxiality Tolerances

It is important to not confuse the concept of concentricity (with reference to geometrical features that have the *same centre*) with coaxiality (with reference to features
that have the *same axis*). In fact, concentricity is the condition whereby an extracted
centre of a circle is congruent with a datum point.

Instead, coaxiality is the condition of a derived median line that should be aligned
with a datum axis. The concentricity tolerance zone, or the coaxiality tolerance zone,
is always circular (concentricity) or cylindrical (coaxiality).

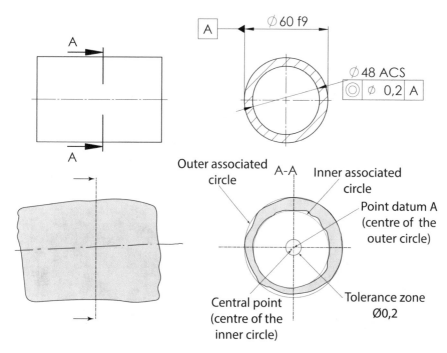

Fig. 9.49 Indications and interpretation of a concentricity tolerance applied to a feature of size, indicated with an ACS (All Cross Section) modifier to point out that the concentricity applies to each transversal section

Figure 9.49 illustrates the indications of a concentricity tolerance applied to a central point of a feature of size, indicated with an **ACS** (*Any Cross Section*) modifier to point out that the concentricity applies to each transversal section. The extracted centre of the inner circle of any cross-section should be within a 0.2 mm diameter circle, which is concentric with datum point A, defined in the same cross-section as the centre of the external associate circle.

A coaxiality tolerance is applied to an extracted median line of a feature of size, as shown in Fig. 9.50. In this case, the extracted median line should fall within a 0.1 mm diameter cylindrical zone, whose axis is datum A, and which is obtained from the cylinder associated with datum feature A.

When the maximum material condition (MMC) is applied to a coaxiality tolerance, the control of a virtual boundary (MMVC) is obtained (Fig. 9.51), with the advantage of obtaining an increase in the tolerance; in this case, the boundary of the hole should not violate the tolerance zone (MMVC) constituted by a 47.9 mm diameter cylinder, whose axis coincides with datum axis A, obtained from the associated external cylinder.

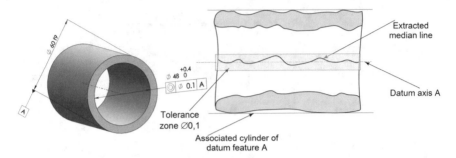

Fig. 9.50 The extracted median line should fall within a 0.1 mm cylinder, whose axis is datum A, obtained from the cylinder associated with datum feature A

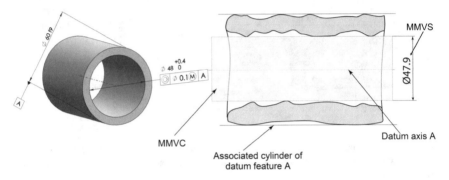

Fig. 9.51 When the maximum material condition (MMC) is applied to a coaxiality tolerance, the boundary of the hole should not violate the tolerance zone constituted by a 47.9 mm cylinder, whose axis coincides with datum axis A, obtained from the associated external cylinder

9.3.1   Concentricity in ASME Y14.5

The concentricity and symmetry symbols **were removed** from the 2018 edition of the ASME Y14.5 standard, thereby eliminating the confusion that surrounds these symbols and their misapplication (many organisations, such as General Motors, had banned the use of these tolerances decades before). These two concepts shown in the 1994 and 1982 versions of Y14.5 have always been controversial and complicated. These symbols controlled the opposing median points of a feature (not the axis or centreplane) relative to a datum and this is rarely a functional requirement.

With respect to the ISO GPS standard, in which a derived median line that should be aligned to a datum axis is controlled, in the case of the ASME standard, if the concentricity tolerance refers to revolution surfaces, the median points of the features diametrically opposite the surfaces should fall within a cylindrical tolerance zone, whose axis coincides with the axis taken as a datum. The diameter of the cylinder is equal to the concentricity control tolerance value (Fig. 9.52). The median points are

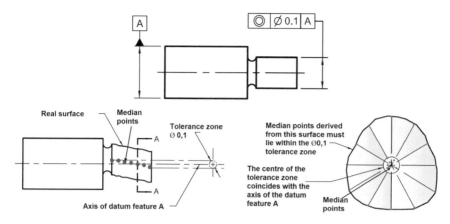

Fig. 9.52 Interpretation of the concentricity tolerance in ASME Y14.5:2009

derived directly from the surface of the workpiece, and the roundness or form errors therefore have an effect on the concentricity tolerance.

In short, since a cloud of points is controlled (not the axis of a feature of size), the maximum material condition and the least material condition are not applicable.

Since inspecting concentricity requires the establishment and verification of the location of the median point of a feature, which must be inside a cylindrical tolerance zone, it requires the inspector to find the median point of many opposite point pairs, thus these tolerances are time consuming and potentially expensive to use. When checking concentricity, all the errors of form of a feature have an effect, since when a dial indicator is in contact with a feature surface, the form errors and coaxiality are indistinguishable from one another and it is necessary to proceed with other methods to control concentricity.

The most typical use of concentricity may involve the verification of the balancing of a rotating part, in spite of the fact that the control is influenced by the inhomogeneity of the material. Although the complexity and difficulty of the verification of concentricity have been demonstrated, American standards advise its use in technical drawings, but prefer the run-out tolerance as an optimal alternative (which controls both the form and coaxiality).

9.4   Symmetry Tolerances

Symmetry represents the condition of a median surface that is congruent with respect to a median plane taken as a datum. A symmetry error is the deviation of the points of the median surface from the plane taken as a datum. Figure 9.53 shows an example of a symmetry tolerance: the extracted median surface should fall between two parallel planes 0.1 mm apart, which are symmetrically arranged about datum plane A.

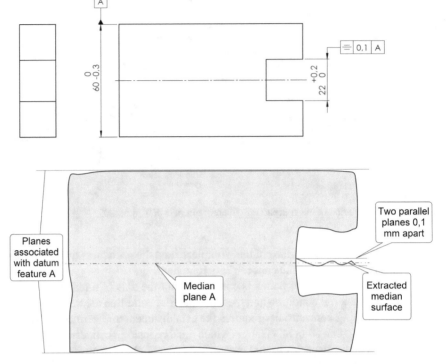

Fig. 9.53 Interpretation of the tolerance zone of a symmetry control. The extracted median surface of the slotting should fall within two parallel planes 0.1 mm apart, arranged symmetrically with respect to the median plane of datum feature A

9.4.1 Symmetry in ASME Y14.5

In the same way as for concentricity, this control has been removed from the ASME standards.

While the error of a derived median surface that should be congruent with a median plane, taken as a datum, is controlled in the ISO standard, in the ASME Y14.5:2009 standard, symmetry represents the condition at which *the median points of all the opposite elements of two surfaces are congruent with respect to a median plane* (or centreplane) taken as a datum. The symmetry error is the deviation of the median points from the plane considered as a datum.

The control of a derived median surface is very complex and costly, and for this reason the indication of this type of tolerance is often discouraged. In order to avoid this problem, the control of symmetry can also be carried out by means of a position tolerance; in this case, the position of an abstract feature, such as a centreplane or axis, is controlled.

Figure 9.54 shows an example of the use of a position tolerance applied to a median plane of a slot and with respect to a datums A and B. The central datum plane, for

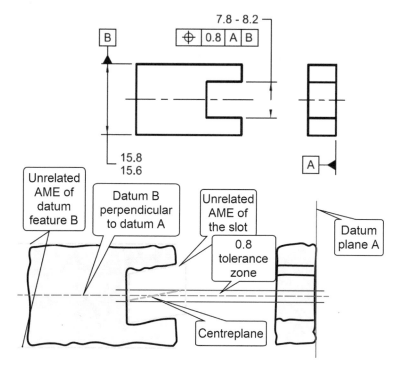

Fig. 9.54 Use of a position tolerance applied to a median plane of a slot

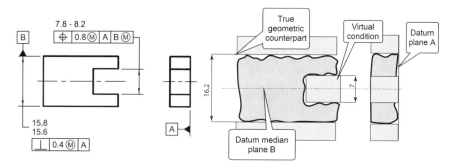

Fig. 9.55 Symmetrical relationship controlled by specifying a positional tolerance at MMC with a datum feature specified on an MMB. No element of the component surfaces should violate a theoretical VC boundary of an identical shape located at the true position

internal features, is constituted by a symmetry plane between two parallel planes which, at the maximum distance, are in contact with the corresponding surfaces of the workpiece. This central plane of the unrelated AME of the slot should remain within a tolerance zone between two parallel planes 0.8 mm apart and should be arranged symmetrically with respect to median plane B.

A symmetrical relationship may be controlled by specifying a positional tolerance at MMC, as in Fig. 9.55. The datum feature may be specified on an MMB, LMB or RMB basis, depending on the design requirements. While specified size limits of the feature of size should be maintained, no element of its surface should violate a theoretical boundary of an identical shape located at the true position.

Chapter 10
Profile Tolerances

Abstract A geometrical tolerance on a profile is one of the most versatile and powerful instruments that can be used for functional dimensioning, and is the tolerance most frequently used by designers. A profile may in fact be used to control the size, form, orientation and location of a feature. Because of the flexibility in the level of control that can be achieved with the profile tolerance, this control may be used to substitute the classical coordinate dimensioning method. The present chapter covers the ISO rules on profile tolerances in order to appropriately specify the control of a profile with a combination of the SZ, CZ, UF symbols and the bidirectional and unidirectional zones that are specified with profile tolerances. A composite profile tolerance is used in the ASME standards when the design applications require stricter tolerances on the form than is needed for the orientation or location of the same feature. The new ASME symbols, that is, From-To and dynamic profile, are presented.

10.1 Introduction

A geometrical tolerance on a profile is one of the most versatile and powerful instruments that can be used for functional dimensioning, in that it is not just a form tolerance, it also controls the **size, orientation, position and, naturally, the form** of a feature. A tolerance on a profile is usually only indicated on complex shaped features, but its use has spread over time to even simple profiles, in such a way that it has become the tolerance *that is used the most frequently by designers.*

The outline of a feature is defined as a **profile**; an outline defined on a drawing by theoretically exact dimensions is called *a theoretical profile.* The control of a profile *specifies the limits, with respect to the theoretical profile, within which the elements of the surface must lie.*

Fig. 10.1 Tolerance symbol
on the profile of a line (on
the left) and of a surface (on
the right)

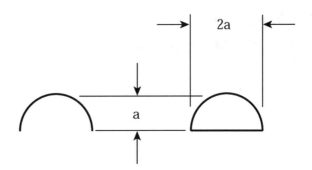

Two methods can be followed to indicate this type of geometrical error on a drawing:

(a) the first indicates the tolerance on the *profile of a line*,[1] such as on an edge or any curve that defines a bi-dimensional tolerance zone; in this case, the symbol that should be placed within the tolerance frame is the one shown on the left in Fig. 10.1. A tolerance on a profile of a line is specified when an error on each feature of the surface becomes critical, and the variation from one feature to another becomes less important.

(b) The second method indicates the *tolerance on the profile of a surface*,[2] in order to obtain the total control of a three-dimensional zone, sometimes with the use of one or more datum features; the symbol in the frame is similar to the previous one, but this time with a closed boundary (symbol on the right in Fig. 10.1).

A control of the profile may be used to substitute the classical coordinate dimensioning method, since tolerance profiles are no more restrictive than coordinate tolerances and are often equivalent. In fact, using a profile is similar to using coordinate tolerances with datums.

A plus/minus location dimension is used in the drawing in Fig. 10.2, since many engineers believe that it is very difficult to check a profile and this approach is more restrictive than coordinate tolerancing. In reality, from the metrologist's point of view, the 90 ± 0.2 mm dimension inspection is a complicated procedure, since, according to ISO 14660-2:1999, the local size of two parallel extracted surfaces is the distance between two points on opposite extracted surfaces, where:

- the connecting lines of sets of opposite points are perpendicular to the associated median plane; and
- the associated median plane is the median plane of two associated parallel planes obtained from the extracted surfaces.

Furthermore, in this approach, the correct distance of the surface from datum A is not checked.

---

[1]ISO: Profile of any line.
[2]ISO: Profile of any surface.

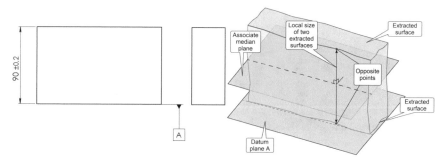

Fig. 10.2 From the metrologist's point of view, the 90 ± 0.2 mm dimension inspection is a compli-cated procedure, since according to ISO 14660-2:1999, the local size of two parallel extracted surfaces is the distance between two points on opposite extracted surfaces, where the connecting lines of sets of opposite points are perpendicular to the associated median plane obtained from the extracted surfaces. Furthermore, in this way, the correct distance of the surface from datum A is not checked

Moreover, as can be seen in Fig. 10.3, the tolerance of ±0.2 on the height of the workpiece is substituted by the tolerance on the profile. The tolerance zone remains unvaried (±0.2) with respect to the theoretically exact size), but the profile control provides a clear definition of the tolerance zone, with uniform boundaries relative to a datum system. The location of the surface, as well as the parallelism, the flatness error and, naturally, the size are also controlled through the tolerance on the profile.

A profile tolerance is very easy to inspect by using a CMM, an optical comparator or a dial indicator, as shown in Fig. 10.4.

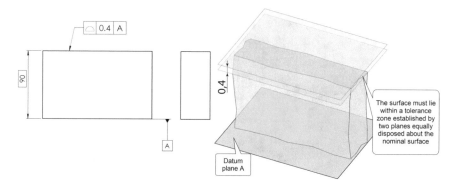

Fig. 10.3 Traditional dimensioning compared with dimensioning of the profile. The tolerance zone remains unvaried (±0.2 with respect to the theoretically correct size), but the profile control provides a clear definition of the tolerance zone

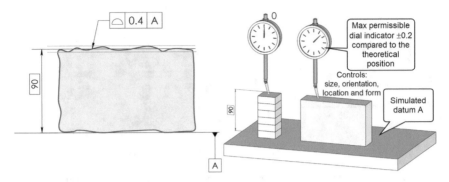

Fig. 10.4   A profile tolerance is very easy to inspect, and the location of the surface, as well as the size, the parallelism and the flatness error are controlled

10.2   Rules for Profile Tolerancing (ISO 1660:2017)

1. When applying a profile tolerance, the **theoretically exact feature** (**TEF**) of the toleranced feature should be defined explicitly with *theoretically exact dimensions* (TEDs) or implicitly with the dimensions embedded in the CAD model.
2. The tolerance zone for line profile specification is limited by two lines that envelope circles with a diameter equal to the tolerance value, the centres of which are situated on the TEF, unless otherwise specified (Fig. 10.5).
3. The tolerance zone for surface profile specification is limited by two surfaces that envelope spheres with a diameter equal to the tolerance value, the centres of which are situated on the TEF (see Fig. 10.6), unless otherwise specified.

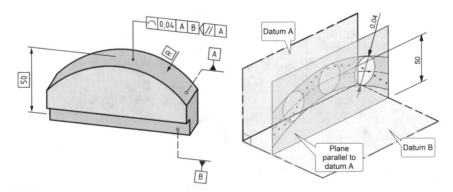

Fig. 10.5   The extracted profile line in each section, parallel to datum plane A, as specified by the intersection plane indicator, should fall between two equidistant lines that envelope 0.04 diameter circles, the centres of which are situated on a line which has the theoretically exact geometrical form with respect to datum plane A and datum plane B

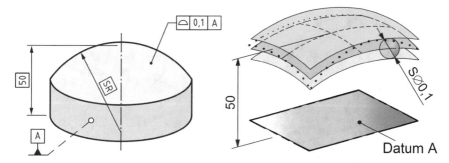

Fig. 10.6 The extracted surface should fall between two equidistant enveloping surfaces with a 0.1 mm diameter sphere, the centres of which are situated on a surface that has the theoretically exact geometrical form with respect to datum plane A

4. According to the feature principle (ISO 8015, 5.4), by default *a profile specification applies to one complete single feature.*
5. According to the independency principle (ISO 8015, 5.5), by default, a profile specification that applies to more than one single feature, applies to those features **independently**. If the profile specification has to apply to a united feature or a combined zone, the appropriate symbology should be specified (Fig. 10.7).
6. The "all over" indication and the "all around" indication *should always be combined with UF, CZ or SZ*, when used for geometrical tolerancing, to make it explicit whether the specification applies to a united feature, defines a combined zone or defines a set of separate zones, except when all the non-redundant degrees of freedom for all the tolerance zones are locked by reference to datums.

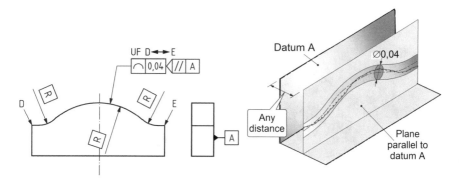

Fig. 10.7 Indication and interpretation of a line profile with an intersection plane indicator, using an intersection plane symbol. According to the independency principle, by default, a profile specification that applies to more than one single feature, applies to those features independently. In this case, it is necessary to use the UF (Unified Feature) symbol to indicate that the three circular sections are combined in one single feature

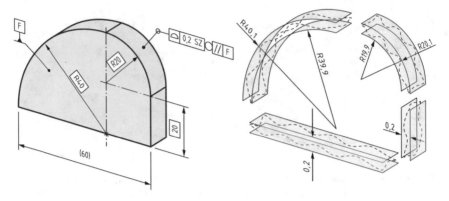

Fig. 10.8 Surface profile specification for a set of independent features

An example of the indication of a surface profile specification for a set of independent features is reported in Fig. 10.8 through the use of a collection plane symbol. To avoid ambiguity, the "all around" symbol is combined with the SZ specification element to indicate that the features are independent.

Since the "all around" symbol is used, the specification applies to a set of features that make up the periphery of the workpiece when seen in a plane parallel to datum F, *as indicated by the collection plane indicator*. The features are considered independent, and the meaning would be the same, if four leader lines were used to identify the four features.

The UF symbol in Fig. 10.9 indicates that the specification applies to a united feature. Since the "all around" symbol and the UF modifier are used, the specification applies to a united feature built from the features that make up the periphery of the

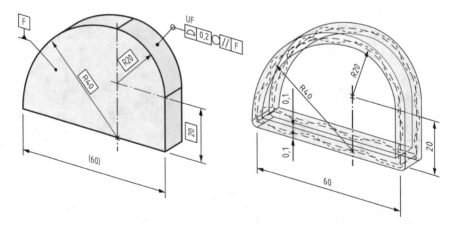

Fig. 10.9 Surface profile specification for a united feature. Since the periphery is considered one feature, the spheres that define the limits of the tolerance zone are rolled across the discontinuities in the feature to create round corners in the tolerance zone on the outside of the discontinuities

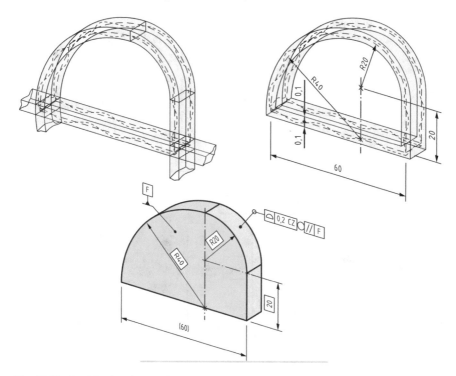

Fig. 10.10 Combined surface profile specification for a set of features. The illustration shows the four zones and how the zones are combined so that the part has sharp outside corners

workpiece *when seen in a plane parallel to datum F,* as indicated by the collection plane indicator. The meaning would have been the same, if four leader lines had been used to identify the four features instead of the "all around" symbol. The specification does not reference datums, and the tolerance zone is therefore not constrained.

The drawing indications in Fig. 10.10 differ from those in Fig. 10.9 in that the CZ modifier is used instead of the UF modifier to indicate that the specification is a combined tolerance zone that applies to a set of features. Since the "all around" symbol is used, the specification applies to a set of features that make up the periphery of the workpiece, when seen in a plane parallel to datum F, as indicated by the collection plane indicator.

10.3 Profile Interpretation

A tolerance on a profile of a surface can be used to control:

- only the size and form of a profile (the profile control is specified without any datum references, Fig. 10.11a);

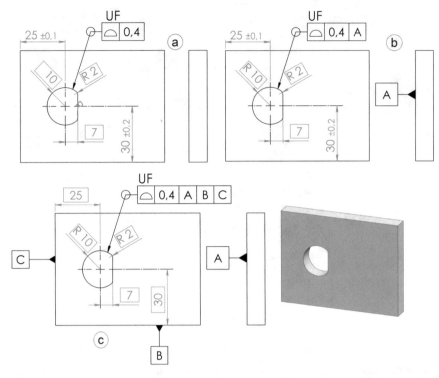

Fig. 10.11 Control of the size and form (**a**), the size, form and orientation (**b**) and the size, form, orientation and position (**c**). In this case, because all the non-redundant degrees of freedom for the tolerance zone are locked by means of reference to a datum system, the UF modifier could have been omitted without changing the practical meaning of the specification

- the size, form and the orientation, through the use of one or more datums (Fig. 10.11b)
- the size, form, orientation and position at the same time (Fig. 10.11c). Because all the non-redundant degrees of freedom for the tolerance zone are locked by means of reference to a datum system, the UF modifier could be omitted without changing the practical meaning of the specification. In this case, the tolerance zone is completely constrained by the datum system.

10.4 Offset Tolerance Zone Specification

By default, the tolerance zone is symmetrical with respect to the ideal profile, but
it is also possible to indicate a non-symmetrical zone through the use of modifiers
(Fig. 10.12). In fact, the ISO 1101 standard has introduced the UZ modifier, with
which the number after this symbol represents the deviation required by the nominal,
and it can therefore be either negative (within the ideal profile) or positive (external).

In practice, the extracted surface should fall between two equidistant enveloping
surface spheres of a defined diameter that is equal to the tolerance value, the centres
of which are situated on a surface that corresponds to the envelope of a sphere in
contact with the theoretically exact feature (TEF) and whose diameter is equal to the
absolute value given after UZ, with the direction of the offset indicated by a sign,
where the "+" sign indicates "outside the material" and the "−" sign indicates "inside
the material" (Fig. 10.13, the position of this sphere depends on whether the sign is
positive or negative). *The offset sign should always be indicated.*

The UZ modifier in Fig. 10.14 is used to control a surface location with the position
symbol. This combination can only be used for planar features.

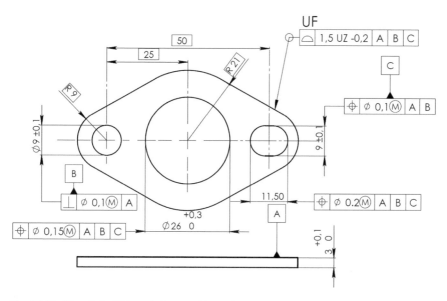

Fig. 10.12 Use of the UZ symbol to specify a tolerance on a non-symmetrical profile. The UF
modifier could be omitted

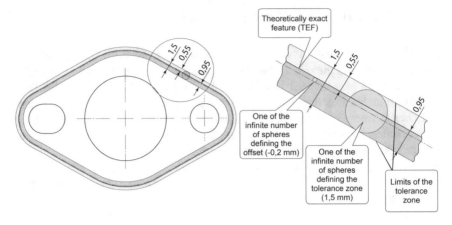

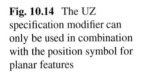

Fig. 10.13 Offset tolerance zone with the specified offset

Fig. 10.14 The UZ
specification modifier can
only be used in combination
with the position symbol for
planar features

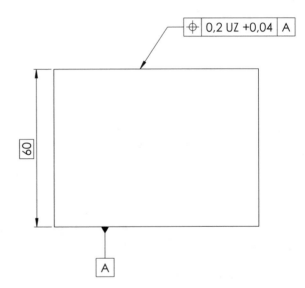

10.5   Constraint Specification Elements

10.5.1   Offset Tolerance Zone with an Unspecified Linear Offset

Another interesting constraint pertaining to the tolerance of a profile, which was
introduced with the ISO 1101 standard of 2017, is the OZ symbol (which should
not be confused with the previous UZ symbol). The OZ symbol has the objective of
indicating a tolerance zone that is offset by a constant but *unspecified amount*. The

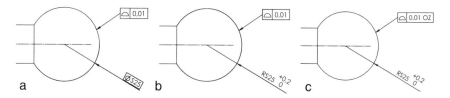

Fig. 10.15 Profile control on a spherical surface. The TEF of the toleranced feature should be defined with TEDs (**a**). The specification in (**b**) is ambiguous, since the nominal size of the TEF is not defined by a TED and there is only a ± tolerance indicated for the size. In this case, the specification element OZ should always be indicated (**c**)

theoretically exact feature (TEF) of the toleranced feature should usually be defined with theoretically exact dimensions (TEDs) in the specification of a profile control, as in Fig. 10.15a. In the case of the sphere in Fig. 10.15b, the specification of *the profile control is ambiguous,* since the nominal size of the TEF is not defined by a TED, and there is only a ± tolerance indicated for the size. In this case, the specification element OZ should always be indicated for profile specifications to make it explicit that **the size of the TEF is not fixed**. In short, if the shape of the TEF is defined, but the nominal linear size of the TEF is undefined, *the OZ modifier should always be indicated.*

Since there are no bounds on the offset, a specification with the modifier is usually combined with a specification using a larger tolerance without a modifier, as in Fig. 10.16. When both specifications are satisfied, this combination controls the shape of the tolerance feature within the larger, fixed tolerance zone.

The non-redundant degrees of freedom for the tolerance zone defined by the upper tolerance indicator are locked by means of reference to a datum system, and the UF modifier could therefore have been omitted. In the case of the tolerance zone defined by the lower tolerance indicator, the UF modifier is necessary to synchronise the offset of the two parts of the tolerance zone.

Fig. 10.16 Combination of a fixed and an off-set specification

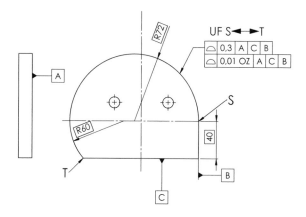

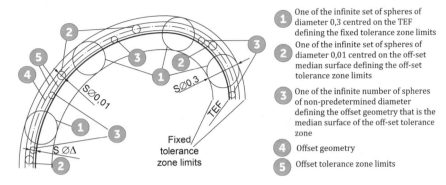

Fig. 10.17 Interpretation of the tolerance zones

Figure 10.17 shows the interpretation of the drawing in Fig. 10.16, where a combination of a fixed tolerance zone with a larger tolerance value (0.3) and an off-set tolerance zone with a smaller tolerance value (0.01) are defined. The off-set tolerance zone is off-set by *a non-predetermined amount*, either inside the material or outside the material, compared to the TEF. The combination of the two specifications controls the shape of the toleranced feature within the off-set tolerance zone, which can adapt to the toleranced feature, as long as the off-set tolerance zone remains within the fixed tolerance zone.

10.5.2   Offset Tolerance Zone with an Unspecified Angular Offset

In the case of a feature of angular size, as in Fig. 10.18, a profile specification of a cone surface is indicated with its size considered as fixed, while the angle of the cone is indicated with a theoretically exact dimension (TED). Therefore, the extracted surface of the cone is required to be inside the tolerance zone, without any orientation or location constraints, according to ISO 3040:2016. The tolerance zone consists of the space between two coaxial conical surfaces 0.3 mm apart, with a specified theoretical angle. However, when a tolerance zone does not take the nominal size into account, the **VA** modifier (*variable angular size*) should be indicated.

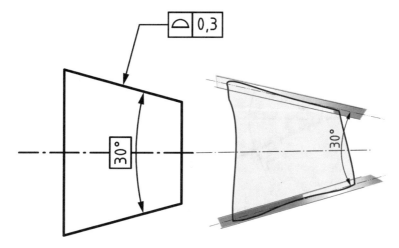

Fig. 10.18 Indication of a profile specification of a cone surface with its size considered as fixed

When a nominal angular size of the TEF is not defined by a TED for cones, e.g. in the case where there is only a ± tolerance indicated for the angular size, the VA specification element should always be indicated for profile specifications to clearly indicate that the angular size of the TEF is not fixed. The extracted surface of the cone in Fig. 10.19 is required to be inside the tolerance zone, without any orientation or location constraints. The tolerance zone consists of the space between two coaxial conical surfaces 0.3 mm apart, with the same unspecified angle. The local angles are required to fall between the lower and upper tolerances.

Since there are no bounds on the angular offset, a specification with the VA modifier is usually combined with another specification such as an angular dimensional specification or geometrical specification, without any VA modifier, as in Figs. 10.19 and 10.20.

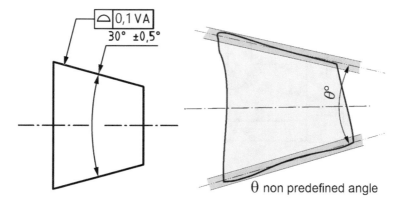

Fig. 10.19 Indication of a profile specification of a cone surface with its size considered as variable

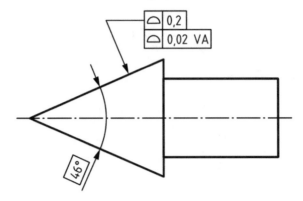

Fig. 10.20 Combination of a fixed and a variable angle specification

10.6   Pattern Specification (ISO 5458:2018)

The ISO 5458:2018 standard establishes rules that are complementary to ISO 1101, which should be applied to pattern specifications, and defines rules to combine individual specifications for the control of profiles. Figure 10.21 provides an example which illustrate the internal constraints introduced by the CZ or CZR modifiers and the external constraints introduced by a datum or datum system. Figure 10.22

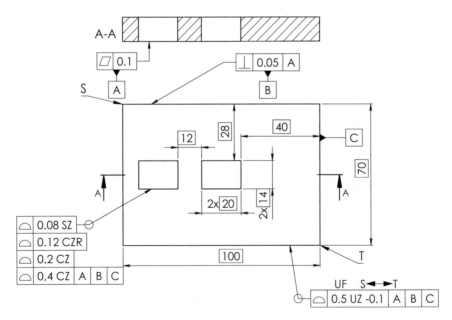

Fig. 10.21 Examples of the internal constraints introduced by the CZ and CZR modifiers and the external constraints introduced by the datum system

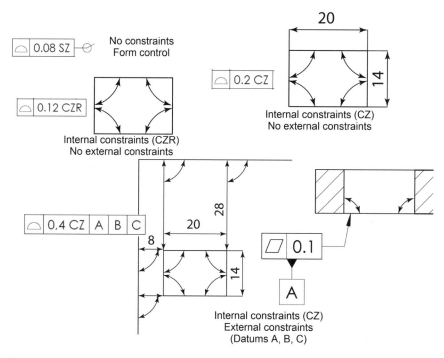

Fig. 10.22 Interpretation of the internal constraints introduced by the CZ or CZR modifiers and the external constraints introduced by a datum or datum system

shows the interpretation of each tolerance indicator. The CZ modifier indicates that a tolerance zone pattern is defined with internal orientation and location constraints between the individual tolerance zones. The CZR modifier indicates that a tolerance zone pattern is defined with internal orientation constraints between the individual tolerance zones.

The specifications in Fig. 10.23 are two pattern specifications (CZ in the SZ CZ sequence) considered independently (first SZ in the sequence). The toleranced feature of each pattern specification is the collection of four extracted integral surfaces (all around symbol), and the tolerance zone pattern is composed of four tolerances zones, constrained to each other in orientation (implicit TEDs 0° and 90°) and in location to be 20 mm (in one direction) and 14 mm (in another perpendicular direction) apart, with explicit TEDs, without any external constraint from a datum system. The two tolerance zone patterns are independent of each other, i.e. they are free to move and rotate in relation to each other.

The specification in Fig. 10.24 is a pattern specification (first CZ in the CZ CZ sequence) defined by two (2 ×) tolerance zone patterns (last CZ in the sequence).

Here, there are two (2x) pattern specifications (last CZ) which are dependent on each other (first CZ), thereby creating *a global pattern specification*. The toleranced

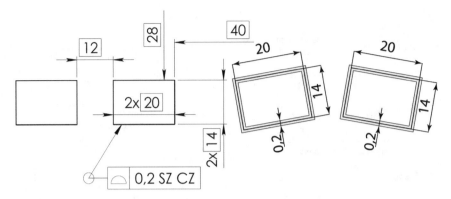

Fig. 10.23 Two pattern specifications (CZ in the SZ CZ sequence) considered independently (first SZ in the sequence). The two tolerance zone patterns are free to move and rotate in relation to each other

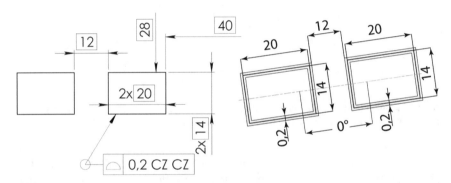

Fig. 10.24 The specification is a pattern specification (first CZ in the CZ CZ sequence) defined by two (2x) tolerance zone patterns (last CZ in the sequence)

feature is the collection of eight extracted integral surfaces (2 × and the *all around* symbol).

The tolerance zone is *a tolerance zone pattern* consisting of two tolerance zone patterns (CZ CZ), composed of four tolerance zones, in a space between two parallel planes 0.2 mm apart and constrained in orientation (implicit TEDs 4 × 90°) and in location (explicit TEDs 20 × 14). The two tolerance zone patterns are constrained in orientation to be parallel (implicit TED of 0°) and in location to be 12 mm apart in one direction (explicit TED) and aligned (implicit TED of 0 mm apart) in the perpendicular direction, without any external constraint from a datum.

NOTE: The last CZ modifier in the CZ CZ sequence creates a tolerance zone pattern composed of four tolerance zones. The first CZ in the CZ CZ sequence creates the dependency between the two tolerance zone patterns.

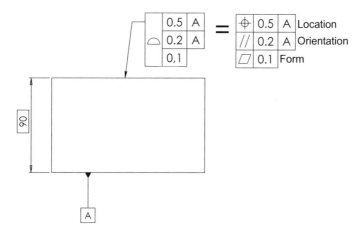

Fig. 10.25 Use of a tolerance on a composite profile. The first uppermost frame controls the position of the surfaces, while the second frame is used for the orientation tolerance. The form tolerance (flatness) and the size are defined by adding a third frame (without adding a datum)

10.7 The Tolerance of a Profile in the ASME Standards

10.7.1 Composite Profile Tolerancing

The ASME Y14.5 standard also foresees the use of **composite profile tolerancing** for profiles (this is not foreseen in the ISO standards and it is also used for location tolerances), a fact which determines a multiple control of the geometric profile, where the design requirements are such that the tolerance for the location of a part feature is less important than the orientation of the feature (Fig. 10.25). The upper segment of a composite profile feature control frame controls the location of the surface, while the second frame is used to control the orientation of a toleranced feature relative to datum A (from necessity, the value must be smaller). By adding a third segment (without adding a datum), a form tolerance is defined (a flatness tolerance, in this case) and the size is also controlled.

In short, in many cases, it becomes more important to control the form and the orientation of a profile than its position, as in the case of the workpiece shown in Fig. 10.26.

10.7.2 Unilateral and Unequally Disposed Profile Tolerance

In order to indicate a tolerance on a non-symmetrical profile, the ASME Y14.5 standard of 2009 introduced the modifier Ⓤ (*unequally* disposed profile symbol) to indicate a unilateral and unequally disposed profile with respect to a theoretically

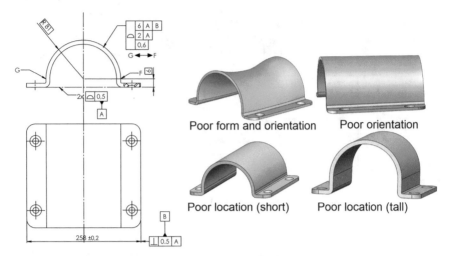

Fig. 10.26 The use of composite profile tolerancing allows the position of a surface to be controlled with a rather large error (first frame, 6 mm) and, at the same time, a more rigorous control of the orientation (2 mm), form and size (0.6 mm) to be obtained

exact profile, as shown in Fig. 10.27. The Ⓤ symbol is placed in the feature control frame after the tolerance value. The value after the symbol indicates the entity of the profile tolerance in the direction *that would allow additional material to be added to the true profile* (if the number is zero, the error is completely inside the profile).

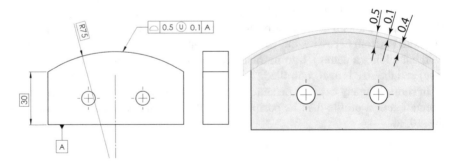

Fig. 10.27 The Ⓤ modifier indicates an unequally disposed profile with respect to a theoretically exact profile. The value after the modifier indicates the entity of the tolerance in the direction that would allow additional material to be obtained

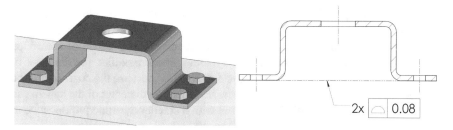

Fig. 10.28 In the ASME standard, a profile of a surface may be used to control the mutual orientation and location of two or more surfaces

10.7.3   Coplanarity

Coplanarity *is the condition of two or more surfaces having all the elements on one plane.* When it is important to treat two or more surfaces as a single continuous one, (a coplanar one, for example), the ISO standards use the CZ (Combined Zone) symbol. In the ASME standard, a profile of a surface tolerance may be used to control the mutual orientation and location of two or more surfaces, as shown in Fig. 10.28. The profile of a surface tolerance establishes two tolerance zones, parallel to each other with zero offset, within which the considered surfaces should lie.

10.7.4   The New Symbols (ASME Y14.5:2018)

The function of the dynamic profile is to allow form to be controlled independently of size. It is a small triangular symbol Δ that can be inserted inside the feature control frame of a profile control (after the tolerance value, as in Fig. 10.29). The tolerance of the profile is usually applied to the theoretical or ideal profile (indicated with

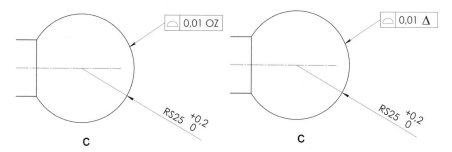

Fig. 10.29 The function of the dynamic profile is to allow form to be controlled independently of size. The concept of dynamic profile corresponds to the OZ (Unspecified linear tolerance zone offset) symbol of the ISO 1101 standard

basic dimensions), for which the shape and size are checked: through the use of the dynamic profile modifier, the error of form is defined independently of the size.

For example, let us suppose that the spherical end of a pin should be controlled by the profile of a surface. Without the dynamic profile modifier, the diameter of the sphere should be indicated with a basic dimension, because the control must be applied to an "ideal profile". With this new modifier, the diameter of the sphere can still have its dimensional tolerance ±, but with a smaller dynamic profile value to control the shape.

When the dynamic tolerance modifier is applied to a lower segment of a composite tolerance without any datum feature references, the tolerance zone controls the form but not the size of the feature and it uniformly progresses (expands or contracts) normal to the true profile while maintaining the specified constant width, as in Fig. 10.30. The 0.5 profile tolerance zone is constrained in translation and rotation relative to the datum reference frame established by datum features A, B, and C. The 0.1 dynamic profile is free to translate, rotate and uniformly progress (expand or contract) and the actual feature should simultaneously be within both tolerance zones.

The new ASME Y14.5:2018 *From-To* symbol indicates a specification transition from one location to another location to define a non-uniform profile tolerance zone. The "from" and "to" locations may be points, lines, or even features, and the leader from the feature control frame should be directed towards the portion of the feature to which that tolerance applies.

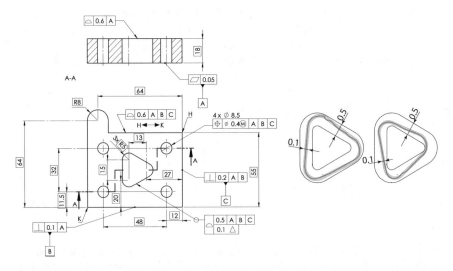

Fig. 10.30 Composite profile with a dynamic profile to control form. The 0.1 dynamic profile is free to translate, rotate and uniformly progress (expand or contract) and the feature should be located within the 0.5 mm tolerance zone

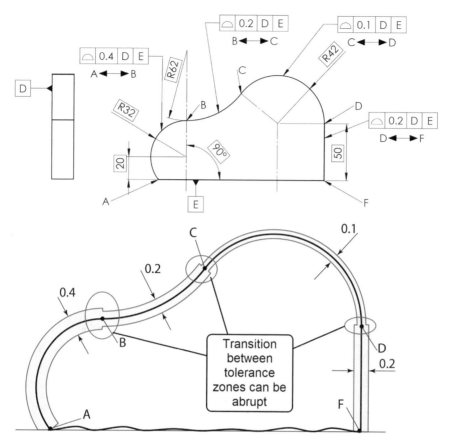

Fig. 10.31 Non-uniform profile tolerance zone with three abrupt tolerance transitions that occur at points B, C, and D, when different profile tolerances are specified

Figure 10.31 shows three abrupt tolerance transitions that occur at points B, C, and D, when different profile tolerances are specified on adjoining segments of a feature. In order to avoid this problem, the *From-To* symbol is used to smooth the transition areas, as in Fig. 10.32.

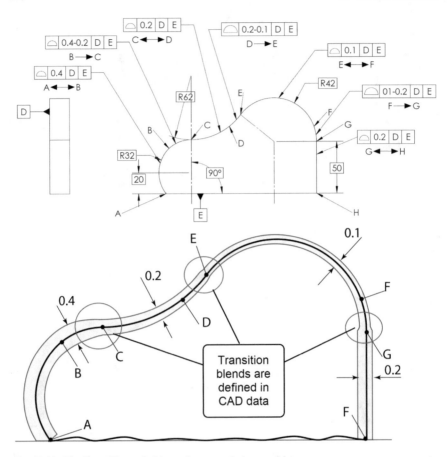

Fig. 10.32 The From-To symbol is used to smooth the transition areas

Chapter 11
Run-Out Tolerances

Abstract A run-out error is the surface variation that occurs relative to a rotation axis. There are two types of run-out controls: circular run-out (2D) and total run-out (3D). Run-out tolerances are composite controls that define the requirements for the permissible coaxiality, orientation and form deviations of a surface element in relation to a datum axis. The application of the run-out control to an assembly and its combination with the tangent plane symbol, are illustrated in this chapter as main novelties of the new ASME Y14.5-2018 standard.

11.1   Introduction

Run-out tolerances are composite controls that define the requirement for the permissible coaxiality, orientation and form deviations of a surface element in relation to a datum axis. The term run-out (runout in the ASME standards) means a deviation of the form, position and orientation of a surface feature during a rotation of the feature itself around a datum axis. In short, a run-out tolerance value specified in a tolerance indicator frame indicates the maximum permissible dial indicator reading of the considered feature, when the part is rotated 360° about its datum axis.

Two types of run-out control exist, that is, **circular run-out** and **total run-out**: the respective symbols are indicated in Fig. 11.1.

Figure 11.2 shows the surfaces that are controlled by this type of tolerance, that is, revolution and perpendicular surfaces, or at least inclined with respect to the axis of the workpiece (surface facing, shoulders). The inspection is generally achieved by means of a dial indicator: in fact, the oscillations of the dial indicator needle first gave rise to the name and symbol, even though the identification is currently achieved through the use of electronic instruments and direct visualisation of the numerical values of the deviations.

Figure 11.3 illustrates the difference between the two types of tolerance and shows the interpretation of the control carried out by means of a dial indicator; in the case of circular run-out, each circular feature of a surface is controlled independently, while the workpiece rotates around the datum axis. Circular run-out provides a 2D control of the circular elements of a surface relative to the datum axis, and a tolerance zone

Fig. 11.1 The symbols used to indicate run-out tolerances: **a** circular run-out e **b** total run-out

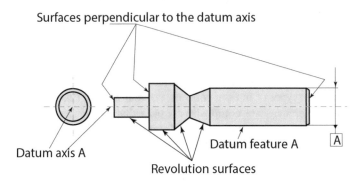

Fig. 11.2 Surfaces controlled by run-out tolerances

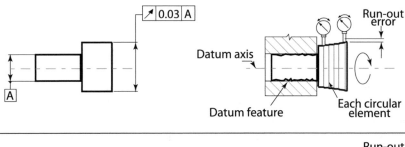

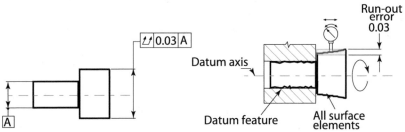

Fig. 11.3 Control of circular run-out and total run-out tolerances: circular run-out applies to each circular element independently, and each circular element must be within its 2D tolerance zone as the part is rotated 360°. A total run-out applies to the entire surface simultaneously, and all the points on the surface must be within a common 3D tolerance zone as the part is rotated 360°

is created and *applied independently to each circular cross section* of the surface of the part.

In the total run-out case, the control applies *to the entire surface simultaneously*; the part is rotated by 360° and the dial indicator is moved along the surface that has to be measured. In this case, the run-out error represents the difference between the maximum and minimum error indicated on the dial indicator at any position of the dial indicator. Total run-out provides a 3D control of all the surface elements of the considered feature and the tolerance applies simultaneously to all circular and profile elements of the surface.

The two run-out controls are both surface controls, and the MMR and LMR requirements *are therefore not applicable.*

The datum feature in run-out tolerances may be constituted by a single axis, two axes considered to be coincident or an axis plus a plane perpendicular to it.

11.2 Circular Run-Out

The control of a circular run-out may be achieved in both a radial and in an axial direction, and the tolerance zone is bi-dimensional. In a radial control, the extracted line, in each transversal section, perpendicular to datum axis A, should fall within two concentric, coplanar circles whose distance is equal to the tolerance; the centre of these circles should be on the datum axis. A circular run-out *applies to each circular element independently*; as already mentioned, the tolerance is applied independently to each measurement position of the workpiece subjected to a rotation of 360°, that is, each individual circular feature should have the circular run-out within the prescribed tolerance.

Figure 11.4 shows an example of circular run-out indications; the tolerance zone is limited, in each plane perpendicular to the axis, by two concentric circles placed at a distance that is equal to the prefixed tolerance and whose centres coincide with the datum axis. Sometimes referred to as **TIR** (*Total Indicated Reading*), run-out is the radial difference between two concentric circles at a datum point, and is designed to include all the points of the measured profile. In the case shown in Fig. 11.4, the radial run-out should be no greater than 0,05 mm on each measurement plane, during a complete rotation around datum axis A.

A circular tolerance zone may also be axial, and therefore allow the control of surfaces perpendicular to the axis, as shown in Fig. 11.5: the extracted line in each cylindrical section, whose axis coincides with that of datum axis D, should fall within two circumferences that are axially 0.1 mm apart. In this case, a circular run-out also controls circular variations (wobble) of a plane surface.

A circular run-out can take place in any direction, or in a specified direction. In the former case, that is, when the direction is not specified, the measurement is performed perpendicular to the surface and the width of the tolerance zone is *normal* to the specified geometry, unless otherwise indicated, as shown in Fig. 11.6. Instead, if the circular run-out specification is indicated in a specified direction, the run-out

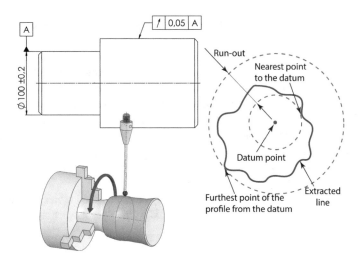

Fig. 11.4 Interpretation and inspection of a run-out tolerance. The extracted line, in each transversal section, perpendicular to datum axis A, should fall within two concentric, coplanar circles, whose distance is equal to the tolerance; the centre of these circles should be on the datum axis

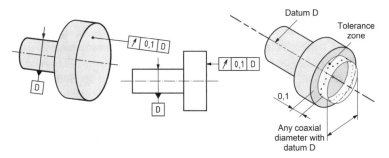

Fig. 11.5 Interpretation of an axial run-out: the tolerance zone is limited in any cylindrical section by two circles, at an axial distance of 0.1 mm apart, lying within the cylindrical section, the axis of which coincides with the datum

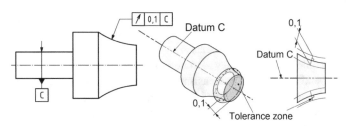

Fig. 11.6 The tolerance zone is limited within any conical section by two circles 0.1mm apart, the centres of which coincide with the datum. The width of the tolerance zone is always normal to the specified geometry, unless otherwise indicated

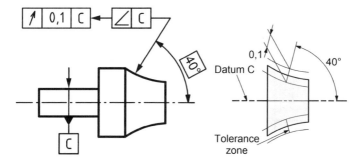

Fig. 11.7 Use of a "*direction feature*" symbol to indicate the measurement direction of the circular run-out error. The tolerance zone is limited within any conical section of the specified angle by two circles 0.1 mm apart, the centres of which coincide with the datum

refers to the specified direction. In this case, the use of a "*Direction feature*" symbol becomes essential, as shown in Fig. 11.7.

11.3 Total Run-Out

In this case, a three-dimensional error zone is controlled and therefore, in the case of radial run-out, the extracted surface should fall within two coaxial cylinders, with a difference in radii equal to the run-out tolerance, the axes of which coincide with the datum axis (Fig. 11.8).

Figure 11.9 shows an example of the control of a tubular tolerance zone determined by two cylinders, and controlled by a complete displacement of the dial indicator (helicoidal movement), while the workpiece rotates around axis A; the difference between the maximum and minimum values shown on the dial indicator represents the total run-out error.

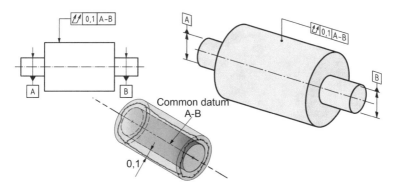

Fig. 11.8 The extracted surface should fall between two coaxial cylinders, with a difference in radii of 0.1 mm, and with the axes coincident with the common datum straight line A–B

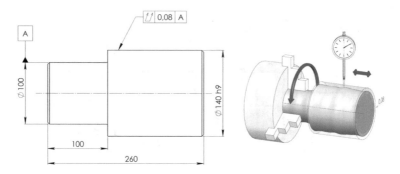

Fig. 11.9 Control of the total run-out with a dial indicator, which is moved in a helicoidal manner

It is also possible to control the total axial run-out in the same way as for the workpiece shown in Fig. 11.10; the extracted surface should fall within two parallel planes (at a distance that is equal to the tolerance) and perpendicular to datum axis D.

The run-out tolerance may also be controlled with two datums, as shown in Fig. 11.11. In this case, the sequence of indications of the datums becomes important: the component is controlled with a self-centring device, which is first placed on datum B and the jaws are then tightened.

Figure 11.12 shows a circular run-out control and a total run-out control through the total movement of a set of dial indicators on each round feature of the surface, while the workpiece rotates around the datum axis. The circular run-out error is obtained in the section with the largest variation (0.06 mm, less the assigned tolerance). The total run-out is the difference between the maximum reading (+0.05 mm) and the minimum reading (−0.08 mm) over the entire surface, that is, 0.13 mm, which is less than the run-out indicated in the design phase.

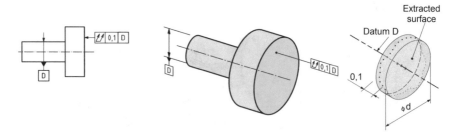

Fig. 11.10 Indications of a total axial run-out: the extracted surface should fall between two parallel planes 0.1 mm apart, which are perpendicular to datum axis D. A perpendicularity specification would have the same meaning

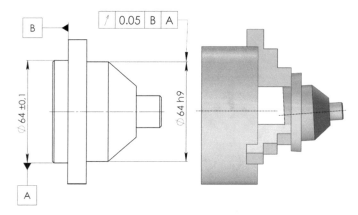

Fig. 11.11 A run-out tolerance may also be controlled with two datums: the component is controlled by means of a self-centring device, which is first placed on datum B and the jaws are then tightened

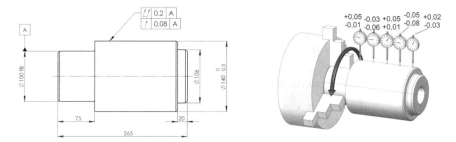

Fig. 11.12 An example of the contemporaneous control of a circular run-out and a total run-out with a set of dial indicators. The total run-out is the difference between the maximum reading (+0.05 mm) and the minimum reading (–0.08 mm) over the entire surface, that is, 0.13 mm, which is less than the run-out indicated in the design phase

11.4 Run-Out Control in the ASME Y14.5 Standards

In the Y14.5:2009 editions, run-out tolerances were explained in terms of a measurement method using *a dial indicator*. For consistency with the explanation method used for other tolerances, the explanation of run-out in ASME Y14.5:2018 *is now based on the resulting tolerance zone*. However, the definition of run-out tolerances has not changed.

- **Y14.5 2009** (Fig. 11.13): *"Total runout provides control of all surface elements. The tolerance is applied simultaneously to all circular and profile measuring positions as the part is rotated 360° about the datum axis. When verifying total runout, the indicator is fixed in orientation normal to and translates along the toleranced surface. The entire surface must lie within the specified runout tolerance zone (0.02 full indicator movement) when the part is rotated 360° about the datum axis*

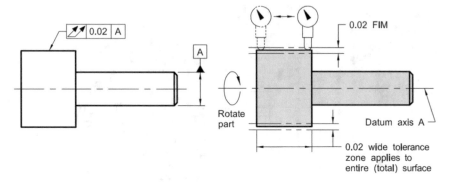

Fig. 11.13 In the Y14.5:2009 editions, run-out tolerances were explained by means of a measurement method using a dial indicator

with the indicator placed at every location along the surface in a position normal to the true geometric shape, without resetting of the indicator. The feature must be within the specified limits of size."

- **Y14.5 2018** (Fig. 11.14): "*Total Runout Tolerance Zone for Cylindrical Features: all surface elements shall be within a tolerance zone consisting of two coaxial cylinders with a radial separation equal to the tolerance value specified. The tolerance zone is constrained in translation (coaxial) to the datum axis.*"

In the ASME standards, the size tolerance controls the size as well as the form of the feature, and Rule #1 requires the perfect form at MMC. Run-out and size tolerances have combined effects on the size, form, orientation, and location of the toleranced feature (Fig. 11.15). Run-out tolerance and size tolerance values are based on the design requirements, and there is no requirement for run-out to be larger or smaller than the size tolerance.

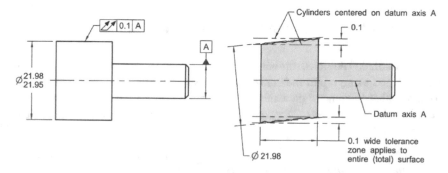

Fig. 11.14 In the new ASME Y14.5:2018 standard, the explanation of run-out is now based on the resulting tolerance zone

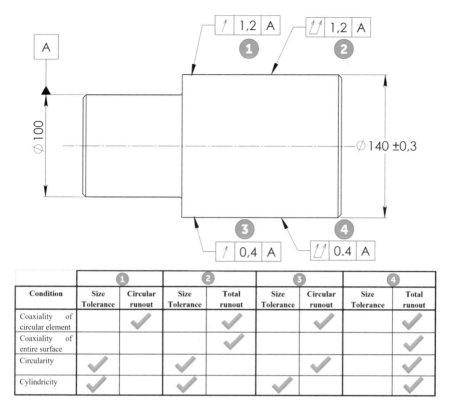

Condition	① Size Tolerance	Circular runout	② Size Tolerance	Total runout	③ Size Tolerance	Circular runout	④ Size Tolerance	Total runout
Coaxiality of circular element		✓		✓		✓		✓
Coaxiality of entire surface				✓				✓
Circularity	✓		✓		✓		✓	
Cylindricity	✓		✓		✓		✓	

Fig. 11.15   Size and run-out tolerance effects

11.4.1   *Run-Out on a Tangent Plane*

The application of a run-out tolerance to a tangent plane has been added to the ASME Y14.5:2018 standard. A run-out tolerance may be applied to a tangent plane for one or more coplanar feature faces that are perpendicular to the rotation axis (Fig. 11.16).

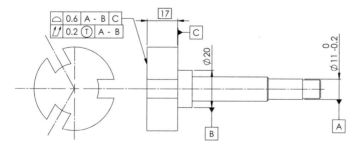

Fig. 11.16   Tangent Plane modifier applied to a total run-out control

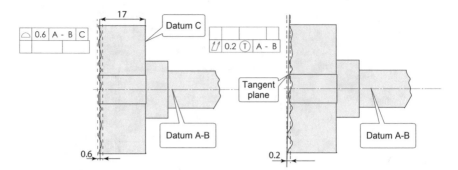

Fig. 11.17 The tolerance zone of the profile is located with reference to datum plane C and oriented to datum A–B, while the tangent plane, in contact with high points of the feature, should be within the 0.2 mm tolerance zone wide perpendicular to datum A–B

The requirements may also be specified using profile tolerances.

The extent of the tangent plane is circular with a radius equal to the distance from the rotation axis to the furthest point on the toleranced surface. In Fig. 11.17, the profile tolerance zone is located with reference to datum plane C and oriented to datum A–B, while the tangent plane, in contact with the high points of the feature, should be within the 0.2 mm wide tolerance zone perpendicular to datum A–B.

11.4.2   Run-Out Tolerance Application to an Assembly

The application of a run-out tolerance relative to an rotation axis, where the axis *may not be a datum feature in an assembly,* has been added to the ASME Y14.5:2018 standard. When specified at an assembly level, run-out requirements *may reference datum features that locate the assembly.* The run-out tolerance is related to the rotation axis, while the axis is constrained at the basic angle relative to the datum reference frame, and the assembly is constrained in translation and rotation to the datum reference frame.

In the example in Fig. 11.18, the datum features constrain the orientation and location of the axis in order to ensure that the toleranced feature is within the specified run-out tolerance of the assembly. The interpretation of the run-out controls is illustrated in Fig. 11.19.

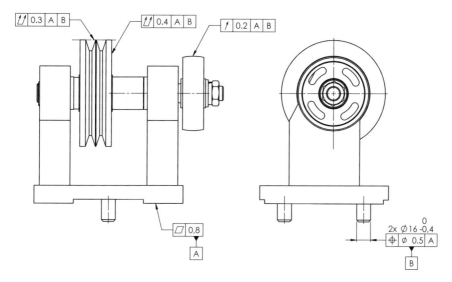

Fig. 11.18   Run-out tolerance applied to an assembly

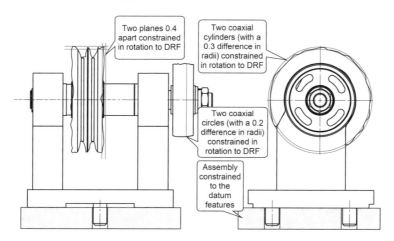

Fig. 11.19   Interpretation of the run-out controls of an assembly

Chapter 12
Geometrical Specification for Non-Rigid Parts

Abstract In the absence of other indications, all the specified dimensions and tolerances are applied and controlled without the action of any force, except gravity, but
components such as thin metal and flexible parts, rubber gaskets, and flexible parts in
general, can deform or bend, even as just a result of their weight. These components
should be inspected under a restrained condition in order to simulate their shape in the
installed condition. The restraint or force on the non-rigid parts is usually applied in
such a manner as to resemble or approximate the functional or mating requirements.
According to the ISO standards, the non-rigid parts should be identified on a drawing
by means of "ISO 10579-NR" (Non Rigid) obligatory indications within or near the
title block; such a note indicates that the part is not considered rigid and it indicates
the clamping forces or other requirements necessary to simulate the assembly conditions. In order to invoke a restrained condition in the ASME standards, a general note,
or a local note, should be specified or referenced on the drawing to define the restraint
requirements. When a general note invokes a restrained condition, all the dimensions
and tolerances apply in the restrained condition, unless they are overridden by a "free
state" Ⓕsymbol

12.1 Non-Rigid Parts

In the absence of other indications, all the specified dimensions and tolerances are
applied and controlled without the action of any force, except gravity. This condition
is defined as the *free state condition*, and it is defined in the ISO 10579-NR standard
as *the condition of a part subject only to the force of gravity*.

The same standard defines the **non-rigid parts** that *deform to an extent that in the
free state is beyond the dimensional and/or geometrical tolerances on the drawing*.
Parts in this category include both those of inherently rigid material (such as thin
metal parts) and those of inherently flexible material (such as rubber, plastics, etc.)
that can bend, even as just a result of their weight.

© The Author(s), under exclusive license to Springer Nature Switzerland AG 2021   283
S. Tornincasa, *Technical Drawing for Product Design*, Springer Tracts
in Mechanical Engineering, https://doi.org/10.1007/978-3-030-60854-5_12

However, this definition is not exactly correct, in that components may be designed which, although they do not change form as a result of their own weight, can deform or bend once a force that is congruent with the control and assembly requirements is applied.

As a result, a correct definition of a rigid part and a non-rigid part could be: "*a rigid part is a component that does not deform or bend by a quantity that prevents its functionality under the effect of forces and assembly constraints*". Vice versa, a part could be defined non-rigid or flexible when it deforms, even under the effect of its own weight (e.g. O-rings), or does *not maintain its form when it is assembled in its functional datums* (stampings, sheet-metal parts, brackets, tubes, gaskets, etc.).

12.2 The ISO 10579-NR Standard

In the ISO standard, the non-rigid parts should be identified on a drawing by means of "**ISO 10579-NR**" (**N**on **R**igid) obligatory indications within or near the title block; such a note indicates that the part *is not considered rigid* and it indicates the clamping forces or other requirements necessary to simulate the assembly conditions. As a general rule, the restraining forces should approximate the forces encountered at the final assembly.

The distortion of a non-rigid part should not exceed the distortion that allows the part to be kept within specified tolerances for its verification, or assembled by applying pressure or forces that do not exceed those that can be expected under normal assembly conditions.

A tolerance frame with the symbol Ⓕ appears in Fig. 12.1. This symbol imposes

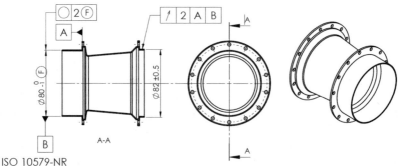

ISO 10579-NR
Restrained condition: the surface indicated as datum A is mounted with 14 M5 bolts (tightened to a torque of 9 Nm to 15 Nm) and the feature indicated as datum B is restrained to the corresponding mating size.

Fig. 12.1 ISO drawing: non-rigid workpieces should be identified on a drawing according to the obligatory ISO 10579-NR standard indications. The geometrical and dimensional tolerance, followed by the Ⓕ symbol, should be controlled at the free state, while the run-out tolerance requires the constraints that are imposed in the note

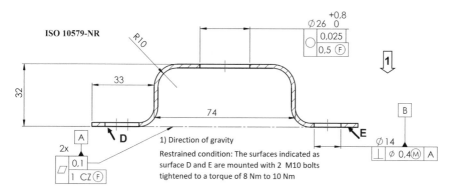

Fig. 12.2 Use of a composite frame to differentiate between the control of a geometrical error at the free state and under a constrained condition

the control of roundness tolerances and dimensional tolerances in the free state, while run-out tolerances are controlled according to the constraint conditions indicated in the note.

Another example is reported in Fig. 12.2, where a composite frame is used to differentiate between the control of a geometrical error at a free state and at a restrained condition.

12.3   Non-Rigid Parts in the ASME Y14.5:2018 Standard

Rule 4.1(n) in the ASME Y14.5:2018 standard establishes that: "*Unless otherwise specified, all dimensions and tolerances apply in a free state condition*". The same ASME standard classifies and defines rigid parts (which do not change form or dimension as a result of their own weight) and non-rigid parts, such as pieces in rubber, thin sheeting or parts in plastic, which undergo a distortion after removal of the forces applied during manufacturing. The distortion of such a part is principally due to the weight and flexibility of the part and the release of internal stresses resulting from fabrication.

There are many parts such as plastic parts, and many types of metal sheeting, that should be inspected under a restrained condition to simulate their shapes in the installed condition. The restraint or force on the non-rigid parts is usually applied in such a manner to resemble or approximate the functional or mating requirements. It may sometimes also be necessary to specify the direction of the gravity force. To invoke a restrained condition, a general note, or a local note, should be specified or referenced on the drawing to define the restraint requirements. When a general note invokes a restrained condition, all the dimensions and tolerances apply under the restrained condition, unless they are overridden by a "free state" Ⓕsymbol (Fig. 12.3).

A geometric tolerance applied to a restrained part or assembly may require one or more free state (unrestrained) datum references. Unrestrained datum features are

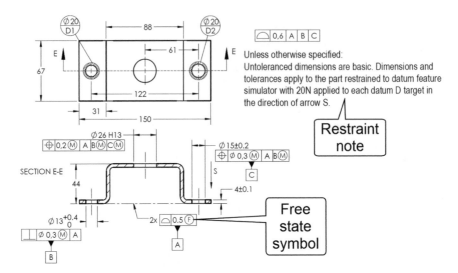

Fig. 12.3 Non-rigid part in the ASME Y14.5 standard. In order to invoke a restrained condition, a general note, or a local note, should be specified or referenced on the drawing to define the restraint requirements. When a general note invokes a restrained condition, all the dimensions and tolerances apply under the restrained condition, unless they are overridden by a "free state" ℰsymbol

designated with the "free state" symbol applied to a datum feature referenced in the feature control frame, following any other modifying symbols that may be present (Fig. 12.4). The datum features or datum targets referenced at a free state for over-constrained datum reference frames on a produced part are not required to be in contact with the physical datum feature simulator when they are in the free state.

Let us take the part shown in Fig. 12.5 as an example [1]. As can be observed, the sheeting structure is made up of two supports anchored to a frame with a series of M6 fasteners. The two supports are mounted back-to-back with two fasteners before they are mounted onto the main structure. A drawing of one of the supports is visible in Fig. 12.6, where, in the absence of other indications, all the tolerances should be verified at the free state, under the action of just gravity.

Figure 12.7 shows the support in the worst manufacturing case, that is, with a convex form and in contact with the centre of datum feature simulator A. A verification of the component under free-state conditions is positive, in that the position tolerance of the holes for the M6 fasteners is respected.

However, when the two bolts are tightened, the mating surfaces are pulled together, and the parts will deform to such an extent that the nominal distance of 92 mm (46 + 46) between each couple of holes *will be greatly reduced*. In these conditions, it is impossible to assemble the part.

In order to avoid this problem, the component is considered *as a non-rigid part*: all non-rigid parts should be restrained under force in order to be controlled, and the applied forces should simulate the expected assembly conditions. Adding a *Restraint Note* to the drawing indicates that the dimensions and tolerances that are applied with

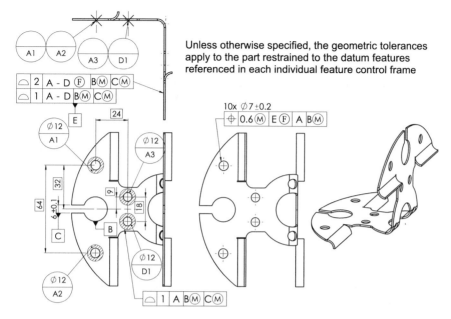

Unless otherwise specified, the geometric tolerances apply to the part restrained to the datum features referenced in each individual feature control frame

Fig. 12.4 Indication of unrestrained datum targets designated with the "free state" Ⓕ symbol

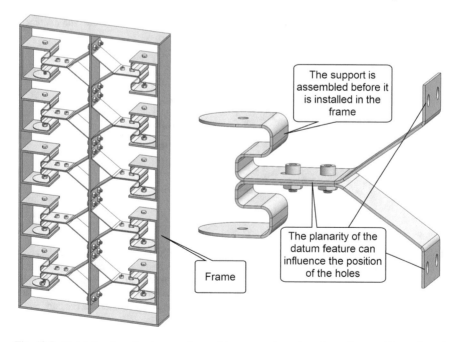

Fig. 12.5 Metal sheeting structure made up of two supports anchored to a frame with a series of M6 fasteners. The two supports are mated with two fasteners before they are assembled to the main structure

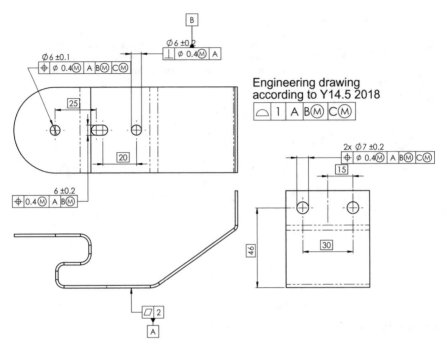

Fig. 12.6 Drawing of one of the supports. In the absence of other indications, all the tolerances should be verified at the free state, under the action of just gravity

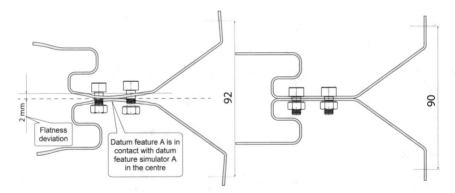

Fig. 12.7 A support in the worst manufacturing state, that is, with a convex form and in contact with the datum feature stimulator. A verification of the component in free-state conditions is positive, in that the position tolerance of the two holes for the fasteners is respected. When the two M6 bolts are clamped, the sheeting deforms in such a way that the nominal distance of 92 mm between each couple of holes is greatly reduced. In these conditions, it is impossible to assemble the part

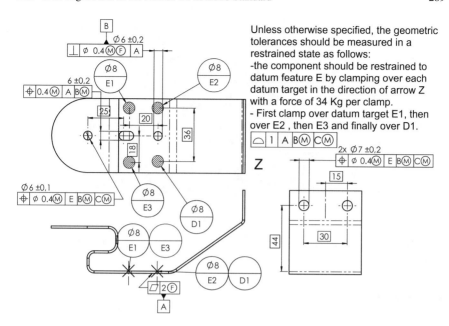

Unless otherwise specified, the geometric tolerances should be measured in a restrained state as follows:

-the component should be restrained to datum feature E by clamping over each datum target in the direction of arrow Z with a force of 34 Kg per clamp.

- First clamp over datum target E1, then over E2 , then E3 and finally over D1.

Fig. 12.8 The dimensioning of a component considered as a non-rigid part: it is necessary to add two additional indications: the use of a restrained note and of a free-state symbol which releases the tolerance from the constraint requirements

forces to the part are approximately the same as those that will be encountered during the assembly process.

The dimensioning is similar to that of rigid parts, but with two additional indications (Fig. 12.8):

(1) The use of a restraint note to indicate the entity of the constraining force, the direction along which the forces are applied and the application sequence.

(2) The use of the free-state symbol (which releases the tolerance from the restraint requirements). In this way, the component is verified with restraints and forces that are congruent with the control and assembly requirements (Fig. 12.9). The flatness tolerance and the position tolerances of the 6 mm hole are controlled with the component at the free state.

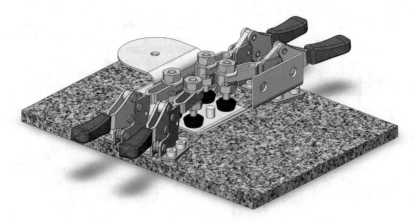

Fig. 12.9 Control procedure with restraints and forces that are congruent with the assembly requirements

Reference

1. Fischer BR (2009) The journeyman's guide to geometric dimensioning and tolerancing: GD&T for the new millennium. Advanced Dimensional Management Press

Chapter 13
Linear Sizes

Abstract In the past, the verification of the local size of workpieces was carried out using traditional metrological instruments, such as callipers and micrometres, or using hard gauges to check for conformance with the global size tolerances. Today, in order to remove any ambiguities that existed in the traditional size tolerance specifications, it is necessary to both exploit the benefits of CMMs and other coordinate measuring systems that collect several points from the surface of a feature and, at the same time, to expand the concept of size tolerancing in order to communicate the function of a part more clearly. The specification operators defined in the ISO 14405 standards provide mechanisms that can be used to expand the domain of size specifications by means of a rich, new set of size modifiers, which provide new capabilities that can be used to address the requirements that arise from many industrial applications.

13.1 Introduction

For many years, the verification of the *local size* of the workpieces was carried out using traditional metrological instruments, such as callipers and micrometres, or using hard gauges to check for conformance to the *global size* tolerances.

In the ASME standards, according to the *Taylor principle* (Rule #1), two separate inspection methods are used to determine conformance to the upper and lower size specifications. The MMC size limit (MMS in ISO) is tested with a *functional gauge* that has the same size as the tolerance limit. If the workpiece is at its maximum material condition (MMC), it cannot undergo any form variation, otherwise it would not fit within the gauge boundary. The LMC size limit (LMS in ISO) is instead inspected using callipers or other devices with opposite measuring faces. It is evident that, if different metrology tools are used, it is possible *to evaluate different pairs of opposite points*, but a contradictory results may arise.

In the ISO GPS standards, the size and form of workpieces *are controlled independently*, and both the upper and lower limits of size are controlled, by default, by two-point measurements. However, again in this case, any measurements carried out

© The Author(s), under exclusive license to Springer Nature Switzerland AG 2021
S. Tornincasa, *Technical Drawing for Product Design*, Springer Tracts
in Mechanical Engineering, https://doi.org/10.1007/978-3-030-60854-5_13

with a calliper depend on the skills and the sensitivity of the metrologist. It is therefore advisable to exploit the benefits of CMMs and/or other coordinate measuring systems that collect several points from the surface of a feature and, at the same time, to expand the concept of size tolerancing to better capture the function of a part.

To meet these needs and to remove any ambiguities that exist in the traditional size tolerance specifications, the latest release of ISO 14405-1:2016 has introduced a rich, new set of size specification modifiers, thereby providing new capabilities to address any requirements that may arise from several industrial applications [1].

13.2   The ISO 14405 Standards: Terms and Definitions

According to the ISO 14405 standards, size is a dimensional parameter, which is considered variable for a feature of size, that can be defined on a *nominal feature* or on an *associated feature*. A size can be *angular* (e.g. the angle of a cone, ISO 14405-3) or *linear* (e.g. the diameter of a cylinder, ISO 14405-1).

The real value of the dimension of a feature of size depend on the form deviations and on the type of specification operator used for the applied size. The type of specification operator, applied to a feature of size, depends on the function of the workpiece, and can be indicated on the drawing by means of *a specification modifier* to control the definition of a feature.

The ISO 14405-1 and ISO 14405-3 standards provide a set of tools that can be used to express several types of size characteristic, defining 16 new symbols to specify linear size tolerances and 11 symbols for angular sizes. Specification modifiers and symbols are shown in Table 13.1, for upper and/or lower limit specification.

Complementary specification modifiers for linear sizes are reported in Table 13.2 together with some indication examples. The combination of these modifiers and symbols is described in Fig. 13.1.

A **local size** is a distance between two opposite points on an extracted integral linear feature of size (e.g. two-point diameter) and *an infinity of local sizes exists* for a given feature. A two-point size is the distance that separates the two points that compose an opposite point pair *taken simultaneously on the extracted feature*. Geometrically speaking, opposite point pairs are obtained from the intersection of a non-ideal integral feature with a*n enabling feature which is a straight line*. In Fig. 13.2, the enabling feature, a straight line, which goes through the centre of the associated circle, allows an opposite point pair to be constructed.

A **global size** is a size characteristic that, by definition, has an *unique* value along and around a toleranced feature of size and it can be evaluated directly (*direct global size*) or indirectly (*indirect global size*, for example, from the average of a set of two-point size values taken on the extracted cylindrical surface).

Table 13.1 Specification modifiers for linear and angular sizes

Modifier	Description
(LP)	Two-point size
(LS)	Local size defined by a sphere
(LC)	Two-line angular size with minimax association criterion
(LG)	Two-line angular size with least squares association criterion
(GG)	Least-squares association criterion
(GX)	Maximum inscribed association criterion
(GN)	Minimum circumscribed association criterion
(GC)	Minimax (Chebyshev) association criteria
(CC)	Circumference diameter (calculated size)
(CA)	Area diameter (calculated size)
(CV)	Volume diameter (calculated size)
(SX)	Maximum linear (angular) size
(SN)	Minimum linear (angular) size
(SA)	Average linear (angular) size
(SM)	Median linear (angular) size
(SD)	Mid-range linear (angular) size
(SR)	Range of linear (angular) sizes
(SQ)	Standard deviation of linear (angular) sizes

Table 13.2 Complementary specification modifiers

Description	Symbol	Example of indication
United feature of size	UF	UF 3x $\varnothing 14 \pm 0.1$ (GN)
Envelope requirement	(E)	$\varnothing 28 \pm 0.2$ (E)
Any restricted portion /Length	/length.	$\varnothing 20 \pm 0.1$ (GG)/15
Any cross section ACS	ACS	$\varnothing 30 \pm 0.1$ (GX) ACS
Specific fixed cross section SCS	SCS	$\varnothing 26 \pm 0.1$ (GX) SCS
Any longitudinal section ALS	ALS	$\varnothing 30 \pm 0.1$ (GX) ALS
More than one feature Number ×	n x	4 x $\varnothing 12 \pm 0.01$ (E)
Common toleranced (FOS) CT	CT	$\varnothing 20 \pm 0.1$(E) CT
Free-state condition	(F)	$\varnothing 30 \pm 0.3$ (F)
Between	◄—►	30 ± 0.3 A ◄—► B
Intersection plane	⟨\\\| B	10 ± 0.01 ALS ⟨⊥ A
Direction feature	◄—\\\| B	10 ± 0.01 ALS ◄⊥ A
Flagnote	⟨1⟩	30 ± 0.2 ⟨1⟩

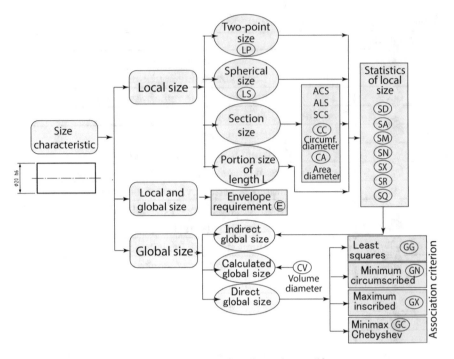

Fig. 13.1 Classification and combination rules of the linear size modifiers

13.3   Default Specification Operator for Size

The ISO default specification operator used for a linear size (without any specification modifier) is the *"two-point size"* LP. If the two-point size (default) is applied for both of the specified limits, the LP modifier should not be indicated.

The ISO default specification operator used for an angular size is the *"two-line angular size"* LC, with a **minimax** association criterion (this requires the use of a CMM). If the two-line angular size is applied for both of the specified limits, the LC modifier can be omitted.

When a drawing-specific default specification operator for size applies, it should be indicated on the drawing in or near the title block by specifying reference to the ISO 14405 standard, and to the general specification modifier(s) chosen for linear and angular sizes (Fig. 13.3). In order to facilitate the reading of the drawing, it is also possible to indicate all of the other types of modifier used on the drawing by listing them in brackets after the default specification indication.

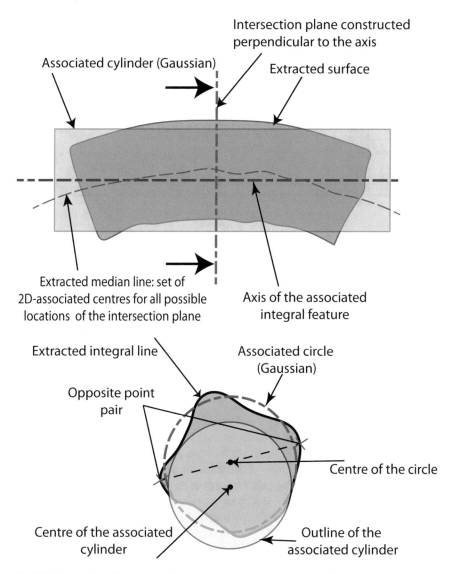

Fig. 13.2 Example of the construction of an opposite point pair on a cylinder

13.4 Drawing Indications

When different specification operators are applied for the upper limit and the lower limit, each specification operator is described using modifiers, even when one is the default operator.

The envelope requirement Ⓔ is a simplified indication that is equivalent to expressing two separate requirements, one for the upper limit of size and another for

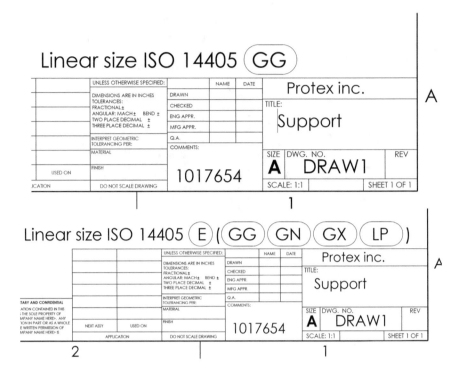

Fig. 13.3 Example of a change in the default specification operator for linear and angular sizes for the entire drawing

the lower limit of size. The GX modifier is used for an internal feature (e.g. the hole in Fig. 13.4), and the GN modifier is used for an external feature (e.g. the shaft in Fig. 13.5) for the maximum material condition of size (upper or lower tolerance). The LP modifier is used for the minimum material condition of the size. It is also possible to add the SX and SN modifiers (maximum and minimum of the set of values of a local size) for the minimum material condition in order to *express the Taylor principle requirement.*

If more than two specification operators are applied to a feature of size, they should be specified by means of the following methods:

- on separate dimension lines (see Fig. 13.6a), each containing one or two specification operators;
- with one-dimension line, attached to several reference lines, each one containing one or two specification operators (Fig. 13.6b);
- with one-dimension line, on which more than one dimensional specification is indicated directly (Fig. 13.6c) and separated by a dash, with each specification being written within square brackets.

Figure 13.7 demonstrates how important it is to communicate the methodology in a precise way through the use of the modifiers indicated in the ISO 14405-1 standard.

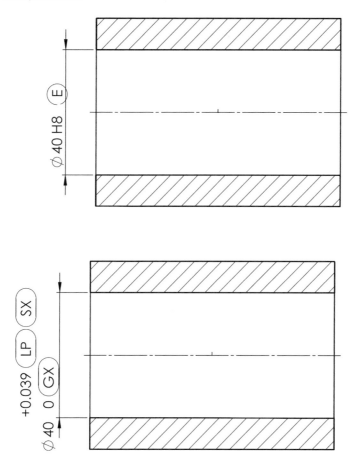

Fig. 13.4 Alternative indications to express the envelope requirement (internal feature)

The specification of size of the 44 mm diameter hole is indicated with four different specification modifiers, which leads to different inspection methods producing a contradictory results.

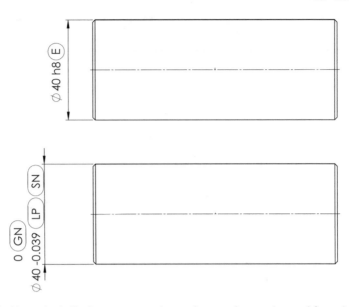

Fig. 13.5 Alternative indications to express the envelope requirement (external feature)

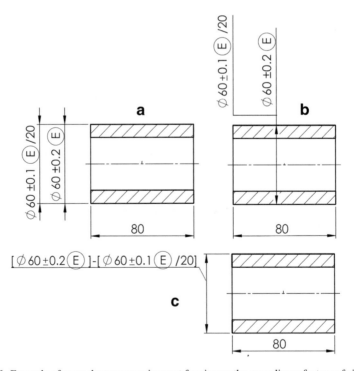

Fig. 13.6 Example of more than one requirement for size on the same linear feature of size

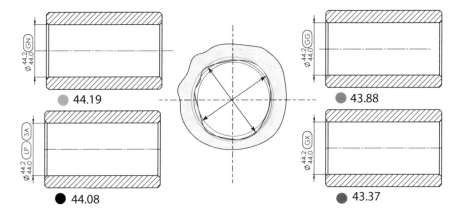

Fig. 13.7 The specification of size of the 44 mm diameter hole is indicated with four different specification modifiers, which leads to different inspection methods producing contradictory results

Reference

1. Morse E, Srinivasan V (2013) Size tolerancing revisited—a basic notion and its evolution in standards. J Eng Manuf 227(5):662–671

Bibliography

Chirone E, Tornincasa S (2018) Disegno tecnico Industriale, vol. I e II, ed. Il capitello
The American Society of Mechanical Engineers (2019) Dimensioning and Tolerancing ASME
    Y14.5:2018

Index

© The Author(s), under exclusive license to Springer Nature Switzerland AG 2021
S. Tornincasa, *Technical Drawing for Product Design*, Springer Tracts
in Mechanical Engineering, https://doi.org/10.1007/978-3-030-60854-5

Printed in the United States
by Baker & Taylor Publisher Services